РАСЧЕТ ЯРКОСТИ СВЕТА В АТМОСФЕРЕ ПРИ АНИЗОТРОПНОМ РАССЕЯНИИ

RASCHET YARKOSTI SVETA V ATMOSFERE PRI ANIZOTROPNOM RASSEYANII

CALCULATION OF THE BRIGHTNESS OF LIGHT IN THE CASE OF ANISOTROPIC SCATTERING

Transactions (Trudy) of the Institute of Atmospheric Physics No. 1

CALCULATION

of the

BRIGHTNESS OF LIGHT

IN THE CASE OF ANISOTROPIC SCATTERING

(Part 1)

E. M. Feigelson, M. S. Malkevich, S. Ya. Kogan, T. D. Koronatova, K. S. Glazova, and M. A. Kuznetsova

TRANSLATED FROM RUSSIAN

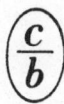

Springer Science+Business Media, B.V.

1960

Original Russian text published by the Academy of
Sciences USSR Press in Moscow in 1958.

Responsible Editor:

Dr. Phys.-Math. Sci., G. B. Rozenberg

ISBN 978-1-4899-5156-4   ISBN 978-1-4899-5154-0 (eBook)
DOI 10.1007/978-1-4899-5154-0

Library of Congress Catalog Card Number 60-8720

TABLE OF CONTENTS

INTRODUCTION

In their monograph [1] E. S. Kuznetsov and B. V. Orchinskii developed methods of approximate solution of the basic equations of the theory of scattering of light in an anisotropically scattering atmosphere and gave some highly accurate results of a numerical solution of the problem for different physical parameters.

Optical characteristics of a real atmosphere change with height and particularly rapidly with time. It is therefore necessary to solve the problem of scattering of light in the atmosphere under more general assumptions as to the laws of scattering than was done in [1].

The present work is looked upon by the authors as an attempt to approximate the real conditions of propagation of light in the terrestrial atmosphere more closely and to discover to what extent the anisotropy in scattering should be taken into account. The present work was carried out at the Institute of Physics of the Atmosphere of the Academy of Sciences of the USSR by members of the staff of the Atmospheric Physics Laboratory.[1] The paper consists of results of calculation of the intensity of solar light scattered by the atmosphere in the case of anisotropic scattering and for different physical parameters and scattering functions. The solution of the integrodifferential equations of the theory of radiation transfer in an anisotropically scattering medium was obtained by the method of successive approximations.[2]

Only part of the calculations, involving the most characteristic cases of the optical state of the atmosphere, is given in the present paper. The second part of this work will contain the remaining material, which includes tables of visibility coefficients and some applications to aerial photography and the visibility range theory.

---

[1] Until the separation of the Institute of Physics of the Atmosphere from the Institute of Geophysics the authors were members of the staff of the Department of Dynamic Meterology, where most of the work was carried out.

[2] The theory was developed by E. S. Kuznetsov [2], [5].

A more general form of the scattering function is considered in the present paper. It is assumed that scattering functions may be expanded into a finite series of Legendre polynomials

$$\gamma(\tau, r', r) = \sum_{i=0}^{N} C_i(\tau) P_i [\cos(r', r)], \tag{1.2}$$

where $C_i(\tau)$ are the expansion coefficients. From the normalization condition

$$\frac{1}{4\pi} \int \gamma \, d\omega' = 1$$

it follows that $C_0 = 1$. Using the addition theorem for Legendre polynomials, we obtain

$$\gamma(\tau, r, r') = \sum_{i=0}^{N} C_i(\tau) \left[P_i(\cos\theta) P_i(\cos\theta') + \right.$$
$$\left. + 2\sum_{m=1}^{i} \frac{(i-m)!}{(i+m)!} P_i^{(m)}(\cos\theta) P_i^{(m)}(\cos\theta') \times \cos m(\psi - \psi') \right], \tag{1.3}$$

where ψ, ψ' are the azimuths of the incident and scattered rays and P_i^m are the associated Legendre polynomials.

We shall give a version of the derivation of the transfer equation for radiant energy in the atmosphere, which was given in [5], for the case where the scattering curve has the form (1.3) but shall alter the boundary conditions so that they correspond to the present problem as formulated above. We shall seek the solution of Eq. (1.1) in the form of a trigonometric series, so that

$$I(\tau, r) = \frac{1}{2} A_0(\tau, \theta) + \sum_{k=1}^{N} A_k(\tau, \theta) \cos k\psi. \tag{1.4}$$

The function $I(\tau, r)$ is even with respect to ψ since the coordinate system was chosen so that the azimuths are measured from the solar vertical.

Let us further denote the upward and downward radiation by

$$I(\tau, r) = \begin{cases} I^{(1)}(\tau, r) & \text{for } 0 \leqslant \theta \leqslant \pi/2, \ 0 \leqslant \psi \leqslant 2\pi \\ I^{(2)}(\tau, r) & \text{for } \pi/2 \leqslant \theta \leqslant \pi \ 0 \leqslant \psi \leqslant 2\pi \end{cases} \tag{1.5}$$

[the expansion coefficients A_k (τ, θ) acquire corresponding indices].

If we substitute (1.3), (1.4), and the analogous expansion for the scattering function in the free term of (1.1) into (1.1), we obtain the following set of equations for $A_k^{(1)}(\tau, \theta)$ and $A_k^{(2)}(\tau, \theta)$:

$$\cos\theta \frac{\partial A_k^{(1)}}{\partial\tau} = \frac{1}{2}\int_0^{\pi/2} A_k^{(1)}(\tau, \theta') \sum_{i=k}^{10} \frac{2}{2i+1} C_i(\tau) \overline{P}_i^{(k)}(\cos\theta)\overline{P}_i^{(k)}(\cos\theta')\sin\theta' d\theta' +$$

$$+ \frac{1}{2}\int_0^{\pi/2} A_k^{(2)}(\tau, \theta') \sum_{i=k}^{10} (-1)^{i-k}\frac{2}{2i+1} C_i(\tau) \overline{P}_i^{(k)}(\cos\theta)\overline{P}_i^{(k)}(\cos\theta')\sin\theta' d\theta' -$$

$$- A_k^{(1)}(\tau, \theta) + \frac{S}{2} e^{-(\tau^*-\tau)\sec\zeta} \sum_{i=k}^{10} (-1)^i \frac{2}{2i+1} C_i(\tau) \overline{P}_i^{(k)}(\cos\theta)\overline{P}_i^{(k)}(\cos\zeta); \quad (1.6)$$

$$- \cos\theta \frac{\partial \dot{A}_k^{(2)}}{\partial\tau} =$$

$$= \frac{1}{2}\int_0^{\pi/2} A_k^{(2)}(\tau, \theta') \sum_{i=k}^{10} (-1)^{i-k}\frac{2}{2i+1} C_i(\tau) \overline{P}_i^{(k)}(\cos\theta)\overline{P}_i^{(k)}(\cos\theta')\sin\theta' d\theta' +$$

$$+ \frac{1}{2}\int_0^{\pi/2} A_k^{(2)}(\tau, \theta') \sum_{i=k}^{10} \frac{2}{2i+1} C_i(\tau) \overline{P}_i^{(k)}(\cos\theta)\overline{P}_i^{(k)}(\cos\theta')\sin\theta' d\theta' - A_k^{(2)}(\tau, \theta) +$$

$$+ \frac{S}{2} e^{-(\tau^*-\tau)\sec\zeta}(-1)^k \sum_{i=k}^{10} \frac{2}{2i+1} C_i(\tau) \overline{P}_i^{(k)}(\cos\theta)\overline{P}_i^{(k)}(\cos\zeta) \quad (1.7)$$

$$(k = 0, 1, 2, \ldots, 10).$$

Eleven terms were taken in the expansion for the scattering function. All the polynomials $P_i^{(k)}$ were normalized (the bar above the symbols which denotes normalization, will be omitted from now on).

Let us now formulate the boundary conditions for the problem. Since the functions $I^{(1)}$ and $I^{(2)}$ represent purely scattered radiation, we shall assume that at the upper boundary of the atmosphere, which corresponds to an optical thickness τ^*,

$$I^{(2)} = 0. \quad (1.8)$$

On the underlying surface ($\tau = 0$), whose reflecting properties are characterized by the albedo $\underline{q}$, the following conditions will be assumed for the reflection of the incident flux of radiation:

CHAPTER I

MATHEMATICAL SOLUTION OF THE PROBLEM

§ 1. Formulation of the Problem. Derivation of Basic Equations

Consider a plane-parallel atmosphere layered horizontally and illuminated at its upper boundary by a parallel beam of solar radiation. It is assumed that at the lower boundary of the atmosphere, i.e., on the earth's surface, the incident radiation is reflected uniformly in all directions (Lambert's law). In that case, the reflected radiation is characterized by the albedo of the earth's surface $\underline{q}$.

With these assumptions the intensity of scattered light $I(\tau, r)$ is given by the following equation:

$$\cos\theta\,\frac{\partial I}{\partial\tau} = \frac{1}{4\pi}\int I(\tau,\,r')\,\gamma(\tau,\,r',\,r)\,d\omega' - I(\tau,\,r) + \\ + \frac{S}{4}\,e^{-(\tau^*-\tau)\,\sec\zeta}\,\gamma(\tau,\,r_{\odot},\,r). \tag{1.1}$$

Here r and r′ are vectors which determine the directions of incidence and scattering;

$$\tau(z) = \int_0^z \sigma(z_1)\,dz_1;$$

τ is the optical thickness of the atmospheric layer $(0, z)$; $\sigma(z)$ is the scattering coefficient; $\tau^* = \tau(\infty)$ is the optical thickness of the whole atmosphere, which depends on transparency P through the relation $P = e^{-T^*}$; $\gamma(\tau, r', r)$ is the scattering function (characteristic curve); $d\omega'$ is a surface element on a unit sphere over which integration is carried out; θ is the angle between the direction of a ray $\underline{r}$ and the vertical axis; πS is the solar constant; ζ is the zenith distance of the sun; and $r_{\odot}$ is the direction of the solar ray.

Equation (1.1) describes the transfer of radiation of wavelength λ.

The scattering coefficient $\sigma(z)$ and the scattering function $\gamma(\tau, r', r)$ are looked upon as given. It is assumed that $\gamma(\tau, r', r)$ possesses axial symmetry. Polarization of light is not taken into account. Scattering will be called spherical or isotropic if for any two directions r and r' the relation $\gamma(\tau, r', r) = 1$ holds. In the case of an arbitrary form of $\gamma(\tau, r', r)$ the scattering will be called anisotropic. Early forms of (1.1) were obtained for the simpler cases of anisotropic scattering (Rayleigh and Rocard functions) in [3], [4].

3

$$F^{(1)}(0) = q F^{(2)}(0), \tag{1.9}$$

where

$$F^{(1)}(0) = \int I^{(1)}(0, r) \cos \theta d\omega;$$

$$F^{(2)}(0) = q \left[\pi S e^{-\tau^* \sec \zeta} \cos \zeta + \int I^{(2)}(0, r) \cos \theta \, d\omega \right].$$

Assuming that Lambert's law holds on the underlying surface, we obtain

$$I^{(1)}(0) = \frac{q}{\pi} \left[\pi S e^{-\tau^* \sec \zeta} \cos \zeta + \int_0^{2\pi} \int_0^{\pi/2} I^{(2)}(0, \theta, \psi) \cos \theta \sin \theta \, d\theta \, d\psi \right]. \tag{1.10}$$

The variation of the scattering function with heigth is taken into account in the simplest way: the atmosphere is divided into two layers in each of which the scattering function is assumed to be independent of height. For this reason, in the set of equations (1.6)-(1.7), instead of a single equation we have two equations for each of the $A_k^{(i)}(\tau, \theta)$, one of which describes the process in the lower layer and the other in the upper one (the dividing boundary will be at τ_1 and functions corresponding to the upper layer will carry a superscript R).

The continuity condition must be obeyed on the surface which separates the two layers, so that

$$I^{(i)}(\tau_1, r) = I^{(i)R}(\tau_1, r) \quad (i = 1, 2). \tag{1.11}$$

From the boundary conditions (1.8)-(1.10) analogous conditions for the function $A_k(\tau, \theta)$ are easily obtained. We shall write them in the following order:

$$A_0^{(1)}(0, \theta) = q \left[\int_0^{\pi/2} A_0^{(2)}(0, \theta) \sin 2\theta d\theta + 2 S e^{-\tau^* \sec \zeta} \cos \zeta \right]; \qquad \left.\begin{array}{c} \\ \\ \end{array}\right\} \tag{1.10a}$$

$$A_k^{(1)}(0, \theta) = 0;$$

$$A_k^{(i)}(\tau_1, \theta) = A_k^{(i)R}(\tau_1, \theta) \quad (i = 1, 2); \tag{1.11a}$$

$$A_k^{(2)R}(\tau^*, \theta) = 0 \tag{1.8a}$$

$$(k = 0, 1, 2, \ldots, 10).$$

It is clear from (1.10a) that the albedo is taken into account only in the determination of the first term of the expansion.

Conditions (1.8a), (1.10a), (1.11a) fully determine the solution of the set of equations (1.6), (1.7). This solution will be obtained below by the method of successive approximations.

We shall determine the functions (i = 1, 2)

$$\overline{A}_k^{(i)} = \frac{A_k^{(i)}}{S/2}\, e^{\tau^* \sec \zeta} \tag{1.12}$$

and the analogous relations for $A_k^{(i)\,R}$, but shall omit the bar above the symbols.

§2. Zero-Order Approximation

In the zero-order approximation the integral terms in (1.6) and (1.7 are omitted. This approximation takes into account only first-order scattering. In all cases, except for a very cloudy atmosphere (large τ^*), this approximation gives a qualitative description of the intensity distribution.

In what follows, this solution will be used to determine the next solution which shall be taken as the starting point of the iteration process.

Let us introduce the following notation

$$F_k^{(1)}(\theta,\, \zeta) = \sum_{i=k}^{10} (-1)^i \frac{2}{2i+1}\, C_i P_i^{(k)}(\cos\theta)\, P_i^{(k)}(\cos\zeta), \tag{1.13}$$

$$F_k^{(2)}(\theta,\, \zeta) = \sum_{i=k}^{10} (-1)^k \frac{2}{2i+1}\, C_i P_i^{(k)}(\cos\theta)\, P_i^{(k)}(\cos\zeta) \tag{1.14}$$

and analogous expressions for the upper layer. In addition, we shall assume that the albedo of the underlying surface is zero, i.e., q = 0. Then, instead of (1.10) we have

$$A_k^{(1)}(0,\, \theta) = 0 \quad (k = 0,\, 1,\, 2,\, \ldots,\, 10). \tag{1.15}$$

In place of the integrodifferential equations (1.6), (1.7) and the boundary conditions (1.8a), (1.11a), (1.15) it is easy to obtain the following equivalent system of integral equations:

$$A_k^{(1)}(\tau,\, \theta) = a_k^{(1)}(\tau,\, \theta) + \frac{1}{2}\sec\theta \int_0^\tau e^{-(\tau-\tau')\sec\theta}\, d\tau' \int_0^{\pi/2} [A_k^{(1)}(\tau',\, \theta')\, R_k^{(1)}(\theta,\, \theta') + \tag{1.16}$$

$$+ A_k^{(2)}(\tau',\, \theta')\, R_k^{(2)}(\theta,\, \theta')]\sin\theta'\, d\theta' \quad (0 \leqslant \tau \leqslant \tau_1);$$

$$A_k^{(2)}(\tau,\ \theta) = a_k^{(2)}(\tau,\ \theta) + \frac{1}{2}\sec\theta \int_{\tau_1}^{\tau^*} e^{-(\tau'-\tau)\sec\theta}\,d\tau' \int_0^{\pi/2} [A_k^{(1)R}(\tau',\ \theta)\,R_k^{(2)R}(\theta,\ \theta') +$$

$$+ A_k^{(2)R}(\tau',\ \theta')\,R_k^{(1)R}(\theta,\ \theta')]\sin\theta'd\theta' +$$

$$+ \frac{1}{2}\sec\theta \int_\tau^{\tau_1} e^{-(\tau'-\tau)\sec\theta}\,d\tau' \int_0^{\pi/2} [A_k^{(1)}(\tau',\ \theta')\,R_k^{(2)}(\theta,\ \theta') +$$

$$+ A_k^{(2)}(\tau',\ \theta')\,R_k^{(1)}(\theta,\ \theta')]\sin\theta'd\theta' \quad (0 \leqslant \tau \leqslant \tau_1);$$

$$\tag{1.17}$$

$$A_k^{(1)R}(\tau,\ \theta) = a_k^{(1)R}(\tau,\ \theta) + \frac{1}{2}\sec\theta \int_0^{\tau_1} e^{-(\tau-\tau')\sec\theta}\,d\tau' \int_0^{\pi/2} [A_k^{(1)}(\tau',\ \theta')\,R_k^{(1)}(\theta,\ \theta') +$$

$$+ A_k^{(2)}(\tau',\ \theta')\,R_k^{(2)}(\theta,\ \theta')]\sin\theta'd'\theta' +$$

$$+ \frac{1}{2}\sec\theta \int_{\tau_1}^{\tau} e^{-(\tau-\tau')\sec\theta}\,d\tau' \int_0^{\pi/2} [A_k^{(1)R}(\tau',\ \theta')\,R_k^{(1)R}(\theta,\ \theta') +$$

$$+ A_k^{(2)R}(\tau',\ \theta')\,R_k^{(2)R}(\theta,\ \theta')]\sin\theta'd\theta' \quad (\tau_1 \leqslant \tau \leqslant \tau^*);$$

$$\tag{1.18}$$

$$A_k^{(2)R}(\tau,\ \theta) = a_k^{(2)R}(\tau,\ \theta) +$$

$$\frac{1}{2}\sec\theta \int_\tau^{\tau^*} e^{-(\tau'-\tau)\sec\theta}\,d\tau' \int_\pi^{\pi/2} [A_k^{(1)R}(\tau',\ \theta')\,R_k^{(2)R}(\theta,\ \theta') +$$

$$+ A_k^{(2)}(\tau',\ \theta')\,R_k^{(1)}(\theta,\ \theta')]\sin\theta'd\theta' \quad (\tau_1 \leqslant \tau \leqslant \tau^*).$$

$$\tag{1.19}$$

Here

$$a_k^{(1)}(\tau,\ \theta) = \frac{e^{\tau\sec\zeta} - e^{-\tau\sec\theta}}{1+\cos\theta\sec\zeta}\,F_k^{(1)}(\theta) \quad (0 \leqslant \tau \leqslant \tau_1); \tag{1.20}$$

$$a_k^{(2)}(\tau,\ \theta) = e^{-(\tau_1-\tau)\sec\theta}\,\frac{e^{-(\tau^*-\tau_1)\sec\zeta} - e^{-(\tau^*-\tau_1)\sec\theta}}{1-\cos\theta\sec\zeta}\,F_k^{(2)R}(\theta)\,e^{\tau^*\sec\zeta}$$

$$(0 \leqslant \tau \leqslant \tau_1); \tag{1.21}$$

$$a_k^{(1)R}(\tau,\ \theta) = e^{-(\tau-\tau_1)\sec\theta}\cdot\frac{e^{\tau_1\sec\zeta} - e^{-\tau_1\sec\theta}}{1+\cos\theta\sec\zeta}\,F_k^{(1)}(\theta) +$$

$$+ \frac{e^{(\tau-\tau_1)\sec\zeta} - e^{-(\tau-\tau_1)\sec\theta}}{1+\cos\theta\sec\zeta}\,e^{\tau_1\sec\zeta}F_k^{(1)R}(\theta) \quad (\tau_1 \leqslant \tau \leqslant \tau^*); \tag{1.22}$$

$$a_k^{(2)R}(\tau, \theta) = \frac{e^{-(\tau^*-\tau)\sec\zeta} - e^{-(\tau^*-\tau)\sec\theta}}{1 - \cos\theta\sec\zeta} e^{\tau^*\sec\zeta} F_k^{(2)R}(\theta) \quad (\tau_1 \leqslant \tau \leqslant \tau^*);$$

$$R_k^{(1)}(\theta, \theta') = \sum_{i=k}^{10} \frac{2}{2i+1} C_i P_i^{(k)}(\cos\theta) P_i^{(k)}(\cos\theta'); \qquad (1.23)$$

$$R_k^{(2)}(\theta, \theta') = \sum_{i=k}^{10} (-1)^{i-k} C_i \frac{2}{2i+1} P_i^{(k)}(\cos\theta) P_i^{(k)}(\cos\theta')$$

and analogous expressions for $R_k^{(i)}R$.

Functions (1.20)-(1.23) considered as zero-order approximations corresponding to the integral equations (1.16)-(1.19) could have been substituted into the corresponding integrals and the next approximation could thus have been found. Since however, the process of integration is very laborious, it is expedient to improve the zero-order approximation from the beginning. The most cumbersome functions in expressions (1.20)-(1.23) are the $F_k^{(i)}$, which are calculated using formulas of the form (1.13), (1.14). The remaining calculations are not too difficult. All the calculations can be carried out on ordinary calculating machines.

Expressions (1.12) and (1.23) lose their meaning for $\theta = \zeta$, since for these values of θ one must use the limiting expressions which are obtained after the indeterminacies are removed:

$$a_k^{(2)}(\tau, \zeta) = [(\tau^*-\tau_1) F_k^{(2)R}(\zeta) + (\tau_1-\tau) F_k^{(2)}(\zeta)] \sec\zeta \cdot e^{\tau\sec\zeta} \quad (0 \leqslant \tau \leqslant \tau_1) (1.21a)$$

$$a_k^{(2)R}(\tau, \zeta) = (\tau^*-\tau) F_k^{(2)R}(\zeta) \sec\zeta \cdot e^{\tau\sec\zeta} \quad (\tau_1 \leqslant \tau \leqslant \tau^*). \qquad (1.23a)$$

§3. Choice of First Approximation

In the first approximation, the effect of multiple scattering on the intensity of the scattered radiation is taken into account in the following way. The required functions A_k are taken out from under the integral sign in Eqs. (1.16)-(1.19) at the point (τ_1, θ_0). If we put

$$A_k^{(i)}(\tau_1, \theta_0) = A_k^{(i)R}(\tau_1, \theta_0) = \tilde{A}_k^{(i)} \quad (i = 1, 2) \qquad (1.24)$$

then from (1.16)-(1.19) it is easy to obtain expressions for the approximation which we shall define as the first approximation.[1]

---
[1] This method of determination of the first approximation is due to I. A. Kibel'.

$$A_{k,1}^{(1)}(\tau,\ \theta) = a_k^{(1)}(\tau,\ \theta) +$$
$$+ [\tilde{A}_k^{(1)}D_k^{(1)}(\theta) + \tilde{A}_k^2 D_k^{(2)}(\theta)] [1 - e^{-\tau \sec \theta}] \quad (0 \leqslant \tau \leqslant \tau_1); \qquad (1.25)$$

$$A_{k,1}^{(1)R}(\tau,\ \theta) = a_k^{(1)R}(\tau,\ \theta) + [\tilde{A}_k^{(1)}D_k^{(1)}(\theta) + \tilde{A}_k^{(2)}D_k^{(2)}(\theta)] [1 - e^{-\tau_1 \sec \theta}] e^{-(\tau-\tau_1)\sec\theta} +$$
$$+ [\tilde{A}_k^{(1)}D_k^{(1)R}(\theta) + \tilde{A}_k^{(2)} D_k^{2)R}(\theta)] [1 - e^{-(\tau-\tau_1)\sec\theta}] \quad (\tau_1 \leqslant \tau \leqslant \tau^*); \ (1.26)$$

$$A_{k,1}^{(2)}(\tau,\ \theta) = a_k^{(2)}(\tau,\ \theta) + [\tilde{A}_k^{(1)}D_k^{(2)R}(\theta) +$$
$$+ \tilde{A}_k^{(2)} D_k^{(1)R}(\theta)] [1 - e^{-(\tau^*-\tau_1)\sec\theta}] e^{-(\tau_1-\tau)\sec\theta} + \qquad (1.27)$$
$$+ [\tilde{A}_k^{(1)}D_k^{(2)}(\theta) + \tilde{A}_k^{(2)}D_k^{(1)}(\theta)] [1 - e^{-(\tau_1-\tau)\sec\theta}] \quad (0 \leqslant \tau \leqslant \tau_1);$$

$$A_{k,1}^{(2)R}(\tau,\ \theta) = a_k^{(2)R}(\tau,\ \theta) + [\tilde{A}_k^{(1)}D_k^{(2)R}(\theta) +$$
$$+ \tilde{A}_k^{(2)}D_k^{(1)R}(\theta)] [1 - e^{-(\tau^*-\tau)\sec\theta}] \quad (\tau_1 \leqslant \tau \leqslant \tau^*). \qquad (1.28)$$

Here we have put

$$D_k^{(i)}(\theta) = \frac{1}{2} \int_0^{\pi/2} R_k^{(i)}(\theta,\ \theta') \sin\theta' d\theta' \quad (i = 1,\ 2), \left.\begin{array}{c} \\ \\ \\ \\ \\ \\ \end{array}\right\}$$
$$D_k^{(i)R}(\theta) = \frac{1}{2} \int_0^{\pi/2} R_k^{(i)R}(\theta,\theta') \sin\theta' d\theta'. \qquad (1.29)$$

The quantities $\tilde{A}_k^{(1)}$, $\tilde{A}_k^{(2)}$ are determined from the set of algebraic equations

$$\tilde{A}_k^{(1)} [1 - M_k^{(1)}] - \tilde{A}_k^{(2)} M_k^{(2)} = a_k^{(1)}(\tau_1,\ \theta_0), \left.\begin{array}{c} \\ \\ \end{array}\right\}$$
$$- \tilde{A}_k^{(1)} M_k^{(2)R} + \tilde{A}_k^{(2)} [1 - M_k^{(1)R}] = a_k^{(2)R}(\tau_1,\ \theta_0). \qquad (1.30)$$

Hence

$$\tilde{A}_k^{(1)} = \frac{a_k^{(1)}(\tau_1,\ \theta_0) [1 - M_k^{(1)R}] + a_k^{(2)R}(\tau_1,\ \theta_0) M_k^{(2)}}{P_k}, \left.\begin{array}{c} \\ \\ \\ \\ \\ \\ \end{array}\right\}$$
$$\tilde{A}_k^{(2)} = \frac{a_k^{(2)R}(\tau_1,\ \theta_0) [1 - M_k^{(1)}] + a_k^{(1)}(\tau_1,\ \theta_0) M_k^{(2)R}}{P_k}, \qquad (1.31)$$

where

$$M_k^{(i)} = D_k^{(i)}(\theta_0) [1 - e^{-\tau_1 \sec \theta_0}] \quad (i = 1,\ 2);$$

$$M_k^{(i)\,R} = D_k^{(i)\,R}(\theta_0)\,[1 - e^{-(\tau^* - \tau_1)\,\sec\theta_0}]\,;$$

$$P_k = [1 - M_k^{(1)}]\,[1 - M_k^{(1)\,R}] - M_k^{(2)\,R}M_k^{(2)}.$$

In the case $\theta = \zeta$ the first approximation is obtained from the same formulas (1.25)-(1.28), except that the expressions for $a_k^{(2)}$, $a^{(2)\,R}$ are taken from the limiting formulas.

Thus in order to obtain the first approximation by the above method, one calculates from the beginning the weighting functions $D_k(\theta)$ which depend only on the form of the scattering function. It is then necessary to choose a point (τ_1, θ_0) at which the unknown functions taken out from under the integral sign are fixed [in our case τ_1 corresponds to the dividing surface between the two layers ($\theta_0 = 45°$)].

The values of A_k^1, A_k^2 depend only on the optical thickness, the form of the scattering function, and the zenith distance of the sun. The obtained approximation could be improved by taking the fixed values of $\widetilde{A}_k^1$, $\widetilde{A}_k^2$ at the intermediate points, for example, $\tau_{1/2}$ ($\tau^* - \tau_{1/2}$) for fixed θ_0, or at the point τ_1 but for somewhat different θ_1 (for example, 30, 60°) or, finally, by taking the values of A_k^1, A_k^2 at all these points. However, this would involve doubling the number of algebraic equations of the form (1.30) in each case, since the number of points at which the unknown functions are fixed is also doubled.

From the point of view of the number of calculations involved, it is expedient to limit ourselves to the above first approximation and to use it to determine the second approximation by the iteration method.

Mathematically, the above method of determination of the first approximation consists of the following. Each of the equations (1.30) for the quantities $\widetilde{A}_k^1$, $\widetilde{A}_k^2$ is an algebraic representation of the corresponding integral equation (1.16)-(1.19) and the required function is determined only at the mean point (τ_1, θ_0). Formulas (1.25)-(1.28) give the usual iteration in which, as the previous approximation to the functions A_k^1, one takes one and the same constant value of A_k^i. The above improvements in accuracy involve the replacement of each integral equation of the set (1.16)-(1.19) by a number of algebraic equations. It is natural to expect that when the magnitudes of the integrals (1.16)-(1.19) are small (this corresponds to small corrections for multiple scattering), the first approximation will give the required intensities of the scattered light to a sufficient accuracy. In the case when the atmosphere is very cloudy (large optical thicknesses and drawn-out scattering curves) more involved corrections must be made for multiple scattering, which in this case is comparable with single scattering, or even exceeds it.

§4. Calculation of Successive Approximations

To obtain the second approximations we substitute expressions (1.25)-(1.28) into the integrals (1.16)-(1.19).

We shall put

$$
\left.
\begin{aligned}
M_k(\theta,\ \theta_s) &= \frac{1}{2}\sum_{i=k}^{10}\frac{2}{2i+1}\,C_i P_i^{(k)}(\cos\theta)\int_{\theta_s}^{\theta_{s+1}} P_i^{(k)}(\cos\theta')\sin\theta'\,d\theta', \\[2mm]
N_k(\theta,\ \theta_s) &= \frac{1}{2}\sum_{i=k}^{10}(-1)^{i-k}\frac{2}{2i+1}\,C_i P_i^{(k)}(\cos\theta)\int_{\theta_s}^{s+1} P_i^{(k)}(\cos\theta')\sin\theta'\,d\theta'
\end{aligned}
\right\} \quad (1.32)
$$

and similarly for M_k^R and N_k^R (instead of C_i we shall have C_i^R). We shall assume in the calculation that

$$
\theta_0 = 0°, \quad \text{with intervals } \theta_{s+1} - \theta_s = 15°, \quad s = 0,\ 1,\ 2,\ 3,\ 4,\ 5.
$$

The inner integrals in (1.16)-(1.19) are replaced by sums (integration by the rectangular method)

$$
\left.
\begin{aligned}
m_k^{(i)}(\tau',\ \theta) &= \sum_{s=0}^{5} A_{k,\,1}^{(i)}(\tau',\ \theta_s)\,M_k(\theta,\ \theta_s), \\[2mm]
n_k^{(i)}(\tau',\ \theta) &= \sum_{s=0}^{5} A_{k,\,1}^{(i)}(\tau',\ \theta_s)\,N_k(\theta,\ \theta_s),
\end{aligned}
\right\} \quad (i=1,\ 2) \quad (1.33)
$$

and similarly for $m_k^{(i)R}$ and $n_k^{(i)R}$.

If we put

$$
\left.
\begin{aligned}
q_k^{(1)}(\tau',\ \theta) &= m_k^{(1)}(\tau',\ \theta) + n_k^{(2)}(\tau',\ \theta), \\[2mm]
q_k^{(2)}(\tau',\ \theta) &= m_k^{(2)}(\tau',\ \theta) + n_k^{(1)}(\tau',\ \theta),
\end{aligned}
\right\} \quad (1.34)
$$

and similarly $q_k^{(i)R}$, then the inner integrals in (1.16)-(1.19) may be written in the form

$$
\left.
\begin{aligned}
Q_k^{(1)}(\tau,\ \theta) &= e^{-\tau\sec\theta}\sec\theta\int_0^{\tau} e^{\tau'\sec\theta}q_k^{(1)}(\tau',\ \theta)\,d\tau' \quad (0\leqslant\tau\leqslant\tau_1), \\[2mm]
Q_k^{(2)}(\tau,\ \theta) &= e^{\tau\sec\theta}\sec\theta\int_{\tau}^{\tau_1} e^{-\tau'\sec\theta}q_k^{(2)}(\tau',\ \theta)\,d\tau' \quad (0\leqslant\tau\leqslant\tau_1), \\[2mm]
Q_k^{(1)R}(\tau,\ \theta) &= e^{-\tau\sec\theta}\sec\theta\int_{\tau_1}^{\tau} e^{\tau'\sec\theta}q_k^{(1)R}(\tau',\ \theta)\,d\tau' \quad (\tau_1\leqslant\tau\leqslant\tau^*), \\[2mm]
Q_k^{(2)R}(\tau,\ \theta) &= e^{\tau\sec\theta}\sec\theta\int_{\tau}^{\tau^*} e^{-\tau'\sec\theta}q_k^{(2)R}(\tau',\ \theta)\,d\tau' \quad (\tau_1\leqslant\tau\leqslant\tau^*).
\end{aligned}
\right\} \quad (1.35)
$$

Consequently, the second approximation is given by

$$
\left.
\begin{aligned}
A_{k,2}^{(1)}(\tau,\ \theta) &= a_k^{(1)}(\tau,\ \theta) + Q_k^{(1)}(\tau,\ \theta), \\
A_{k,2}^{(2)}(\tau,\ \theta) &= a_k^{(2)}(\tau,\ \theta) + e^{-(\tau_1-\tau)\sec\theta}\, Q_k^{(2)R}(\tau_1,\ \theta) + Q_k^{(2)}(\tau,\ \theta), \\
A_{k,2}^{(1)R}(\tau,\ \theta) &= a_k^{(1)R}(\tau,\ \theta) + e^{-(\tau-\tau_1)\sec\theta}\, Q_k^{(1)}(\tau_1,\ \theta) + Q_k^{(1)R}(\tau,\ \theta), \\
A_{k,2}^{(2)R}(\tau,\ \theta) &= a_k^{(1)R}(\tau,\ \theta) + Q_k^{(2)R}(\tau,\ \theta).
\end{aligned}
\right\} \quad (1.36)
$$

The calculation of the second approximation is very laborious. However, as will be shown below, the quantities $A_k^{(i)}(\tau,\ \theta)$ $(i = 1,\ 2)$ fall off rapidly as $\underline{k}$ increases, which means that we can limit our attention to a small number of terms in the series (1.4) and calculate terms corresponding to large $\underline{k}$ with only low relative accuracy. In practice, the second approximation was calculated only for $k = 0,\ 1,\ 2$ and in some case for $k = 3$.

The calculations of the outer integrals in (1.35), as well as the inner ones, was carried out by the rectangular method. For this purpose, integrals (1.35) were represented in the following way. The interval $(0,\ \tau_1)$ was divided into the strips $[\tau^{(1)},\ \tau^{(2)}],\ldots,\ [\tau^{(n-1)},\ \tau^{(n)}]$, $\tau^{(n)} = \tau_1$ and, similarly, the interval (τ_1, τ^*) was divided into strips $[\tau_1, \tau_2]\ldots\ldots[\tau_{n-1}, \tau_n]$ by the points $\tau_1, \tau_2\ldots\tau_n = \tau^*$. Then by taking from under the integrals (1.35) the values of the functions q_k on the left-hand side of each strip for the lower layer $[0, \tau_1]$, and the right-hand side in the case of the upper layer $[\tau_1, \tau^*]$, one can replace each of these integrals by the sums

$$
\left.
\begin{aligned}
Q_k^{(1)}(\tau^{(j)}, \theta) ={}& q_k^{(1)}(\tau^{(1)}, \theta)\, [e^{-(\tau^{(j)}-\tau^{(1)})\sec\theta} - e^{-\tau^{(j)}\sec\theta}] + \\
&+ q_k^{(1)}(\tau^{(2)}, \theta)\, [e^{-(\tau^{(j)}-\tau^{(2)})\sec\theta} - e^{-(\tau^{(j)}-\tau^{(1)})\sec\theta}] + \\
&+ q_k^{(1)}(\tau^{(3)}, \theta)\, [e^{-(\tau^{(j)}-\tau^{(3)})\sec\theta} - e^{-(\tau^{(j)}-\tau^{(2)})\sec\theta}] + \ldots \\
\ldots &+ q_k^{(1)}(\tau^{(j-1)}, \theta)\, [e^{-(\tau^{(j)}-\tau^{(j-1)})\sec\theta} - e^{(\tau^{(j)}-\tau^{(j-2)})\sec\theta}] + \\
&+ q_k^{(1)}(\tau^{(j)}, \theta)\, [1 - e^{-(\tau^{(j)}-\tau^{(j-1)})\sec\theta}], \\
& Q_k^{(1)}(0,\ \theta) = 0;
\end{aligned}
\right\} \quad (1.37)
$$

$$
\left.
\begin{aligned}
Q_k^{(2)}(\tau^{(j)},\ \theta) ={}& q_k^{(2)}(\tau^{(j)},\ \theta)\, [1 - e^{-(\tau^{(j+1)}-\tau^{(j)})\sec\theta}] + \\
&+ q_k^{(2)}(\tau^{(j+1)},\ \theta)\, [e^{-(\tau^{(j+1)}-\tau^{(j)})\sec\theta} - e^{-(\tau^{(j+3)}-\tau^{(j)})\sec\theta}] + \\
&+ q_k^{(2)}(\tau^{(j+2)},\ \theta)\, [e^{-(\tau^{(j+2)}-\tau^{(j)})\sec\theta} - e^{-(\tau^{(j+3)}-\tau^{(j)})\sec\theta}] + \ldots \\
\ldots &+ q_k^{(2)}(\tau^{(n-2)},\ \theta)\, [e^{-(\tau^{(n-2)}-\tau^{(j)})\sec\theta} - e^{-(\tau^{(n-1)}-\tau^{(j)})\sec\theta}] + \\
&+ q_k^{(2)}(\tau^{(n-1)},\ \theta)\, [e^{-(\tau^{(n-1)}-\tau^{(j)})\sec\theta} - e^{-(\tau^{(n)}-\tau^{(j)})\sec\theta}], \\
& Q_n^{(2)}(\tau^{(n)},\ \theta) = 0;
\end{aligned}
\right\} \quad (1.38)
$$

$$
\begin{aligned}
Q_k^{(1)R}(\tau_i,\ \theta) = {}& q_k^{(1)R}(\tau_2,\ \theta)\,[e^{-(\tau_j-\tau_2)\sec\theta} - e^{-(\tau_j-\tau_1)\sec\theta}] + \\
& + q_k^{(1)R}(\tau_3,\ \theta)\,[e^{-(\tau_j-\tau_3)\sec\theta} - e^{-(\tau_j-\tau_2)\sec\theta}] + \\
& + q_k^{(1)R}(\tau_4,\ \theta)\,[e^{-(\tau_j-\tau_4)\sec\theta} - e^{-(\tau_j-\tau_3)\sec\theta}] + \ldots \\
\ldots {}& + q_k^{(1)R}(\tau_{j-1},\ \theta)\,[e^{-(\tau_j-\tau_{j-1})\sec\theta} - e^{-(\tau_j-\tau_{j-2})\sec\theta}] + \\
& + q_k^{(1)R}(\tau_j,\ \theta)\,[1 - e^{-(\tau_j-\tau_{j-1})\sec\theta}], \\
& Q_k^{(1)R}(\tau_1,\ \theta) = 0;
\end{aligned}
\qquad (1.39)
$$

$$
\begin{aligned}
Q_k^{(2)R}(\tau_i,\ \theta) = {}& q_k^{(2)R}(\tau_j,\ \theta)\,[1 - e^{-(\tau_{j+1}-\tau_i)\sec\theta}] + \\
& + q_k^{(2)R}(\tau_{j+1},\ \theta)\,[e^{-(\tau_{j+1}-\tau_j)\sec\theta} - e^{-(\tau_{j+2}-\tau_j)\sec\theta}] + \\
& + q_k^{(2)R}(\tau_{j+2},\ \theta)\,[e^{-(\tau_{i+2}-\tau_i)\sec\theta} - e^{-(\tau_{j+3}-\tau_j)\sec\theta}] + \ldots \\
\ldots {}& q_k^{(2)R}(\tau_{n-2},\ \theta)\,[e^{-(\tau_{n-2}-\tau_j)\sec\theta} - e^{-(\tau_{n-1}-\tau_j)\sec\theta}] + \\
& + q_k^{(2)R}(\tau_{n-1},\ \theta)\,[e^{-(\tau_{n-1}-\tau_j)\sec\theta} - e^{-(\tau_n-\tau_j)\sec\theta}], \\
& Q_k^{(2)R}(\tau^*,\ \theta) = 0.
\end{aligned}
\qquad (1.40)
$$

Consequently, to calculate the outer integral one must first find the weights which are given by the square brackets of expressions (1.37)-(1.40). These weights represent, for each θ, triangular matrices which are calculated once and for all for given τ^*. Formulas (1.36) give the second approximation to $A_k(\tau,\theta)$. Using the weights obtained from (1.37)-(1.40), one can calculate subsequent approximations of any order.

§5. The Effect of Albedo of the Underlying Surface

Next, let us consider the solution of Eqs. (1.6), (1.7) in the case when the albedo of the underlying surface is not equal to zero, i.e., let us take into account the boundary condition (1.10a), which, according to (1.12), has the form

$$
A_0^{(1)}(0,\ \theta) = q\left[\int_0^{\pi/2} A_0^{(2)}(0,\ \theta)\sin 2\theta\,d\theta + 4\cos\zeta\right],
$$

$$
A_k(0,\ \theta) = 0 \quad (k = 1,\ 2,\ \ldots,\ 10).
$$

The set of integral equations which is equivalent to the set of equations (1.6), (1.7) and the boundary conditions (1.8a), (1.10a), (1.11a) will change in form compared with (1.16)-(1.19) only for $A_0^{(1)}(\tau,\theta)$, $A_0^{(1)R}(\tau,\theta)$. The integral equations for the latter quantities have the form

$$\bar{A}_0^{(1)}(\tau,\,\theta) = a_0^{(1)}(\tau,\,\theta) + q\Phi_0 e^{-\tau\sec\theta} +$$

$$+ q e^{-\tau\sec\theta} \int_0^{\pi/2} \sec\theta \left\{ \int_{\tau_1}^{\tau^*} e^{-\tau'\sec\theta} \cdot \frac{1}{2} \left[\int_0^{\pi/2} \bar{A}_0^{(1)\,R}(\tau',\,\theta')\,R_0^{(2)\,R}(\theta,\,\theta')\sin\theta'd\theta' + \right. \right.$$

$$\left. + \int_0^{\pi/2} \bar{A}_0^{(2)\,R}(\tau',\,\theta')\,R_0^{(1)\,R}(\theta,\,\theta')\sin\theta'd\theta' \right] d\tau' +$$

$$+ \int_0^{\tau_1} e^{-\tau'\sec\theta} \cdot \frac{1}{2} \left[\int_0^{\pi/2} \bar{A}_0^{(1)}(\tau',\,\theta')\,R_0^{(2)}(\theta,\,\theta')\sin\theta'd\theta' + \right.$$

$$\left. \left. + \int_0^{\pi/2} \bar{A}_0^{(2)}(\tau',\,\theta')\,R_0^{(1)}(\theta,\,\theta')\sin\theta'd\theta' \right] d\tau' \right\} \sin 2\theta d\theta +$$

$$+ \frac{1}{2}\sec\theta \int_0^{\tau} e^{-(\tau-\tau')\sec\theta} \left[\int_0^{\pi/2} \bar{A}_0^{(1)}(\tau',\,\theta')\,R_0^{(1)}(\theta,\,\theta')\sin\theta'd\theta' + \right.$$

$$\left. + \int_0^{\pi/2} \bar{A}_0^{(2)}(\tau',\,\theta')\,R_0^{(2)}(\theta,\,\theta')\sin\theta'd\theta' \right] d\tau' \quad (0 \leqslant \tau \leqslant \tau_1) \tag{1.41}$$

$$\bar{A}_0^{(1)\,R}(\tau,\,\theta) = a_0^{(1)\,R}(\tau,\,\theta) + q\Phi_0 e^{-\tau\sec\theta} +$$

$$+ q e^{-\tau\sec\theta} \int_0^{\pi/2} \sec\theta \left\{ \int_{\tau_1}^{\tau^*} e^{-\tau'\sec\theta} \cdot \frac{1}{2} \left[\int_0^{\pi/2} \bar{A}_0^{(1)\,R}(\tau',\,\theta')\,R_0^{(2)\,R}(\theta,\,\theta')\sin\theta'd\theta' + \right. \right.$$

$$\left. + \int_0^{\pi/2} \bar{A}_0^{(2)\,R}(\tau',\,\theta')\,R_0^{(1)\,R}(\theta,\,\theta')\sin\theta'd\theta' \right] d\tau' +$$

$$+ \int_0^{\tau_1} e^{-\tau'\sec\theta} \cdot \frac{1}{2} \left[\int_0^{\pi/2} \bar{A}_0^{(1)}(\tau',\,\theta')\,R_0^{(2)}(\theta,\,\theta')\sin\theta'd\theta' + \right.$$

$$\left. \left. + \int_0^{\pi/2} \bar{A}_0^{(2)}(\tau',\,\theta')\,R_0^{(1)}(\theta,\,\theta')\sin\theta'd\theta' \right] d\tau' \right\} \sin 2\theta d\theta +$$

$$+ \frac{1}{2}\sec\theta \int_0^{\tau_1} e^{-(\tau_1-\tau')\sec\theta} \left[\int_0^{\pi/2} \bar{A}_0^{(1)}(\tau',\,\theta')\,R_0^{(1)}(\theta,\,\theta')\sin\theta'd\theta' + \right.$$

$$\left. + \int_0^{\pi/2} \bar{A}_0^{(2)}(\tau',\,\theta')\,R_0^{(2)}(\theta,\,\theta')\sin\theta'd\theta' \right] d\tau' +$$

$$+ \frac{1}{2}\sec\theta \int_{\tau_1}^{\tau} e^{-(\tau-\tau')\sec\theta} \left[\int_0^{\pi/2} \bar{A}_0^{(1)\,R}(\tau',\,\theta')\,R_0^{(1)\,R}(\theta,\,\theta')\sin\theta'd\theta' + \right.$$

$$\left. + \int_0^{\pi/2} \bar{A}_0^{(2)\,R}(\tau',\,\theta')\,R_0^{(2)\,R}(\theta,\,\theta')\sin\theta'd\theta' \right] d\tau' \quad (\tau_1 \leqslant \tau \leqslant \tau^*). \tag{1.42}$$

Here we have put

$$\Phi_0 = \int\limits_0^{\pi/2} a_0^{(2)}(0, \theta) \sin 2\theta d\theta + 4\cos\zeta$$

(the bar above A_0 means that $q \neq 0$).

The form of the integral equations for $A_0^{(2)}(\tau, \theta)$ and $A_0^{(2)R}(\tau, \theta)$ will not change, but the values of these functions will change, compared with the case $q = 0$, because of changes in the values of $A_0^{(1)}$, $A_0^{(2)}$. The remaining A_k ($k = 1, 2, \ldots$) remain unaltered. Consequently, additional calculations are only necessary for $k = 0$. We shall put

$$C = q\Phi_0 + \frac{1}{2} q \int\limits_0^{\pi/2} \sec\theta \left\{ \int\limits_{\tau_1}^{\tau^*} e^{-\tau'\sec\theta} \left[\int\limits_0^{\pi/2} \overline{A}_0^{(1)\,R}(\tau', \theta,) R_0^{(2)\,R}(\theta, \theta') \sin\theta'd\theta' + \right. \right.$$

$$+ \int\limits_0^{\pi/2} \overline{A}_0^{(2)\,R}(\tau', \theta') R_0^{(1)\,R}(\theta, \theta') \sin\theta'd\theta' \bigg] d\tau' +$$

$$\hfill (1.43)$$

$$+ \int\limits_0^{\tau_1} e^{-\tau'\sec\theta} d\tau' \left[\int\limits_0^{\pi/2} \overline{A}_0^{(1)}(\tau', \theta') R_0^{(2)}(\theta, \theta') \sin\theta'd\theta' + \right.$$

$$\left. \left. + \int\limits_0^{\pi/2} \overline{A}_0^{(2)}(\tau', \theta') R_0^{(1)}(\theta, \theta') \sin\theta'd\theta' \right] \right\} \sin 2\theta \, d\theta$$

and

$$A_0^{(i)'} = A_0^{(i)} + CB_0^{(i)} \quad (i = 1,2) \hfill (1.44)$$

and similarly for $A_0^{(1)R}$.

Let us substitute (1.42) into (1.41) (and also into (1.17) and (1.19), where $A_0^{(2)}$, $A_0^{(2)R}$ are replaced by $\overline{A}_0^{(2)}$, $\overline{A}_0^{(2)R}$). We then obtain Eqs. (1.16)-(1.19) for $A_0^{(i)}$, $A_0^{(i)R}$. The functions $B_0^{(i)}$, $B_0^{(i)R}$ are determined from analogous equations, where for $B_0^{(1)}$, $B_0^{(1)R}$ the free term is replaced by the function $e^{-\tau\sec\theta}$, and for $B_0^{(2)}$, $B_0^{(2)R}$ the free term is zero.

The functions $B_0^{(i)}$, $B_0^{(i)R}$ are determined in the same way as the $A_0^{(i)}$, $A_0^{(i)R}$. The following were taken as the zero-order approximation.

$$b_0^{(1)} = b_0^{(1)\,R} = e^{-\tau\sec\theta}, \qquad b_0^{(2)} = b_0^{(2)\,R} = 0.$$

The first approximation is determined from formulas entirely analogous to (1.25)-(1.28), (1.30), (1.31), where

$$\tilde{B}_0^{(1)} = \frac{[1 - M_0^{(1)} R] e^{-\tau_1 \sec \theta_0}}{P_0}, \quad B_0^{(2)} = \frac{M_0^{(2)} R e^{-\tau_1 \sec \theta_0}}{P_0}.$$

The second approximation is determined from formulas analogous to (1.37)-(1.40).

Substituting (1.44) into (1.43) and remembering (1.7) taken at $\tau = 0$, we obtain

$$C = q \frac{\int_0^{\pi/2} A_0^{(2)} (0, \theta) \sin 2\theta \, d\theta + 4 \cos \zeta}{1 - q \int_\pi^{\pi|2} B_0^{(2)} (0, \theta) \sin 2\theta \, d\theta}. \tag{1.45}$$

Thus, using the above relations, the effect of the albedo may be taken into account quite simply. It is easy to see that C depends on the same quantities as $A_0^{(2)}$ (0, θ) and also on $\underline{q}$. To determine the intensities $I_0^{(1)}$ and $I_0^{(2)}$ one must sum up series of the form (1.4) (subscript 0 indicates that q = 0) and then multiply by $\frac{s}{2} e^{-\tau \, \ast \sec \zeta}$. To determine $I_q^{(1)}$ and $I_q^{(2)}$ one adds expressions of the form $CB_0 \frac{s}{4} e^{=\tau \, \ast \sec \zeta}$ to the corresponding intensities for q = 0. More detailed explanations of the calculations are given with the tables (Chap. III).

CHAPTER II

TREATMENT OF OBSERVATIONAL DATA

§1. Review of Observational Data

The development of the theory of scattering of light in the atmosphere is hindered by the lack of information about its optical characteristics, particularly local characteristics, as functions of height. Thus, data on the vertical distribution of the scattering function are at present available only in Waldram's papers [6]. His results are also given in [7] and [8].

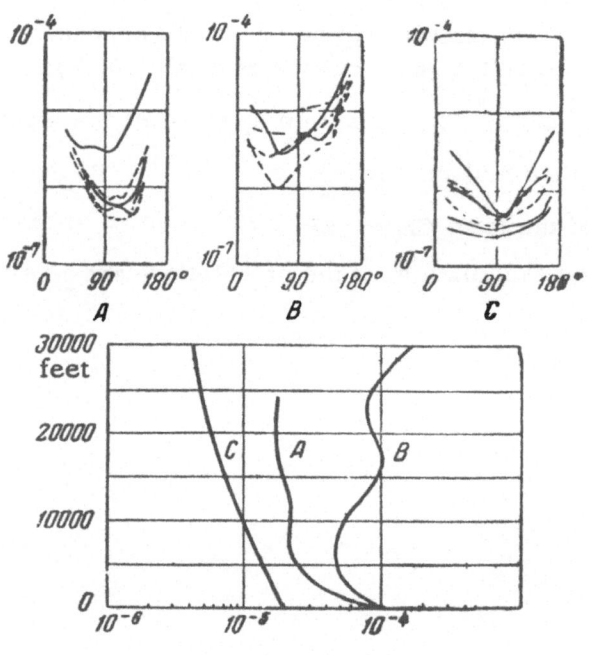

Fig. 1

Waldram carried out night measurements using a nephelometer taken up by an aeroplane to a height of about 10 km. By integrating the scattering function at a given height over all directions, he obtained the scattering coefficient. The dependence of the scattering functions and coefficients on height which were obtained by Waldram for three types of the atmosphere [very transparent (C), cloudy (B), and transparent but misty (A)] are given in Fig. 1.

Measurements of the functions σ (z) and γ (r,r') near the earth's surface have been carried out very intensively. The most widely used method of determination of γ (r,r') near the earth's surface is based on the measurement of the brightness of a horizontal searchlight beam at various angles to its axis.

Measurements of this kind were carried out by V. F. Belov [9], Hulburt [10], Rocard [11], Bullrich [12], Bullrich and Möller [13], Reeger and Seidentopf [14], Foitzik and Zschaeck [15], and others.

The most interesting are the results given in the last three papers because of the very reliable methods employed, the very satisfactory organization of observations,

18

and the amount of collected material. The agreement between them is only general, as can be seen from Fig. 2.[1]

The discrepancy between the curves shown in Fig. 2 may possibly be explained, to some extent, by their classification with respect to horizontal ranges of visibility S_h which are not uniquely related to the scattering functions but are only correlated with them. However, it is natural to try and describe the scattering function by some system of parameters, and, above all, to introduce a main parameter which would be closely related to the scattering function rather than just correlated with it. In many cases it is convenient to take the ratio

$$\frac{\Gamma_1}{\Gamma_2} = \frac{\int_0^{\pi/2} \gamma(\varphi) \sin \varphi \, d\varphi}{\int_{\pi/2}^{\pi} \gamma(\varphi) \sin \varphi \, d\varphi} \qquad (2.1)$$

as the above main parameter. Here, φ is the angle between the incident and the scattered ray. The quantity Γ_1/Γ_2 (Richtungsquotient, defined by Foitzik) represents the characteristic of the scattering function which determines the ratio of the fractions of light scattered into the forward (relative to the incident beam) and backward hemisphere.

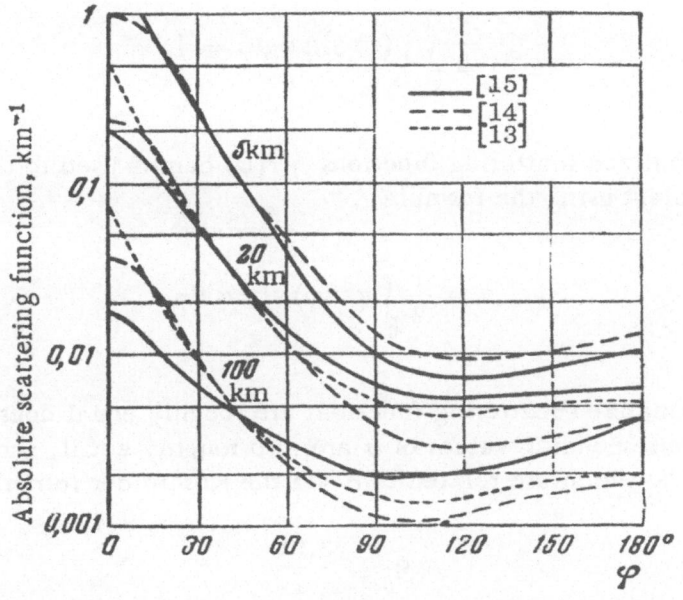

Fig. 2

Table 1 gives examples of scattering functions with close values of Γ_1/Γ_2 obtained by various authors.

[1] V. F. Belov's results are not included since they were only published in June, 1956.

TABLE 1

φ° Author	0	20	40	60	80	100	120	140	160	180
Foitzik and Zschaeck ($\Gamma_1/\Gamma_2 = 8.4$; $S_h = 10$)	15.4	6.5	2.1	0.76	0.35	0.24	0.20	0.20	0.21	0.22
Reeger and Seidentopf ($\Gamma_1/\Gamma_2 = 7.9$; $S_h = 30$)	14.9	7.0	1.9	0.75	0.34	0.22	0.20	0.23	0.26	0.29
Reeger and Seidentopf ($\Gamma_1/\Gamma_2 = 9$; $S_h = 20$)	15.3	7.2	1.9	0.74	0.33	0.20	0.18	0.20	0.23	0.25

Table 1 shows that when the ratios Γ_1/Γ_2 are approximately equal, the scattering functions are also approximately equal at all points. However, the mathematical meaning of the ratio Γ_1/Γ_2 indicates that approximately equal values of it may also correspond to very different values of $\gamma(\varphi)$.

Similar examples will be given below.

The scattering functions given in Table 1 are normalized in accordance with the condition

$$\frac{1}{2} \int_0^\pi \gamma(\varphi) \sin \varphi \, d\varphi = 1. \qquad (2.2)$$

Absolute nonnormalized scattering functions $\gamma^*(\varphi)$ can be used to determine the scattering coefficient using the formula

$$\sigma = \frac{1}{2} \int_0^\pi \gamma^*(\varphi) \sin \varphi \, d\varphi. \qquad (2.3)$$

The fact that normalized scattering functions are roughly equal does not at all indicate that the corresponding values of σ are also roughly equal, and this applies also to the quantities S_h which are related to σ via the Koshmider formula

$$S_h = \frac{3,91}{\sigma}. \qquad (2.4)$$

This is indicated by Table 1 and Fig. 2. Nevertheless, the horizontal range of visibility is related to the form of the scattering functions. There is evidence indicating that, within the limits of a single series of observations, a decrease in the values of S_h, i.e., a reduction in visibility, is accompanied by an increase in the ratio Γ_1/Γ_2, i.e., the extent to which the characteristic curve is drawn out (Table 2).

TABLE 2

Observations of Reeger and Seidentopf

S_h, km	5	10	20	30	40	50	100	347.6
Γ_1/Γ_2	11.2	19.4	9	7.9	6.9	6.3	4.0	1

Observations of Foitzik and Zschaeck

S_h, km	1	2	5	10	20	50	100	200
Γ_1/Γ_2	15.3	15.0	14.0	8.4	4.7	2.5	1.8	1.3

The latter authors give a graph of the dependence of Γ_1/Γ_2 on σ or S_h in [15].

After consideration of the above experimental material on the scattering functions for regions close to the earth's surface we preferred to use the data of Foitzik and Zschaeck. To begin with, they are the most recent and take into account the experience of previous studies. Furthermore, these data are the most reliable in the small-angle region. In some cases direct measurements were carried out down to $\varphi = 2.5°$. Next, Foitzik and Zschaeck, in addition to integral curves, also give spectral curves corresponding to $\lambda = 405/436, 545,$ and 579 mμ, i.e. green, yellow and red light.

These authors [15] quote not only the relative but also the absolute scattering functions, and, together with them, the scattering coefficients. This work contains the largest number of measurements and is divided into characteristic groups, and the mean values of all the main parameters are calculated for each group.

The third type of studies of atmospheric scattering functions is exemplified by the work of E. V. Pyaskovskaya-Fesenkova [16-18] and others. The main method of investigation employed by this author consists in the measurement of the brightness of the sky in the circumsolar halo. From results of measurements she determines the relative scattering function averaged over the whole thickness of the atmosphere. We have used the data of E. V. Pyaskovskaya-Fesenkova in considering a single-layer model of the atmosphere, i.e., a scattering function independent of height. Examples of this kind of calculation will be given in the second part of the present work.

Most of the experimental data in the literature concern the transparency of the atmosphere. These data show that the optical thickness of a cloudless atmosphere lies within the limits 0.1-0.8. The values of τ^* between 0.1 and 0.2 correspond to a very transparent atmosphere, those between 0.2 and 0.4 to a transparent atmosphere, those between 0.4 and 0.6 to cloudy atmosphere, and those between 0.6 and 0.8 to a very cloudy atmosphere.

§ 2. The Use of Experimental Data

In solving Eq. (1.1) we could in choosing the optical characteristics of the atmosphere, i.e., the quantities τ^*, $\sigma(z)$, $\gamma(z,\mathbf{r},\mathbf{r}')$ use only the above experimental material. However, it does not contain all the information necessary to the calculation of the intensity of scattered light in the atmosphere for all the possible optical states of the latter. It was therefore necessary to introduce a number of simplified assumptions about the relative character of the dependence of the scattering functions and the scattering coefficients on height. At the same time, we carried out a tabulation of the values of the intensity for the largest possible number of parameters including, generally speaking, the whole range within which they can lie.

Calculations were carried out for the following values of the parameters:

$$\tau^* = 0.1, \quad 0.2, \quad 0.3, \quad 0.4, \quad 0.6 \quad 0.8;$$
$$\zeta = 30, \quad 45, \quad 60, \quad 75°;$$
$$q = 0.1, \quad 0.2, \quad 0.3, \quad 0.4, \quad 0.6 \quad 0.8.$$

In Eq. (1.1) the height is replaced by the optical thickness τ so that $\sigma(z)$ does not enter into the solution directly. A knowledge of this quantity is only required in the practical application of the solutions when one tries to establish the connection between $\underline{z}$ and τ. How this relation can be established will be indicated below.

The greatest difficulty lies in choosing the scattering function and in taking into account its variation with height. Since we did not have available sufficiently reliable experimental material on the latter dependence, nor a method of solution of Eq. (1.1) for $\gamma(z,\mathbf{r},\mathbf{r}')$ depending on height, we divided the atmosphere into two layers and assigned a scattering function independent of height to each of the layers.[1] The boundary between the layers is taken to be at the point

$$\tau_1 = \frac{3}{4}\tau^*.$$

The upper layer of the atmosphere was assigned a scattering function corresponding to a visibility $S_h = 200$ km and $\Gamma_1/\Gamma_2 = 1.3$, i.e., a very transparent atmosphere. We assumed that the upper layer, which according to the condition $\tau_1 = \frac{3}{4}\tau^*$, begins at a height of about $7 - 10$ km, cannot be considered as free of aerosols, and hence the scattering curve should not be very drawn out but, nevertheless, different from a Rayleigh curve.

In the lower layer of the atmosphere the scattering function was assumed to be identical with one of the functions given by Foitzik and Zschaeck, corresponding to $S_h = 10, 20, 33, 50$, and 100 km, or $\Gamma_1/\Gamma_2 = 1.8, 2.5, 4.7$, and 8.4. Since there exists a sufficiently close correlation between the visibility S_h and the transparency of the atmosphere or the optical thickness τ^*, the calculation was carried out for combinations of the parameters given in Table 3.

Scattering functions II, III, IV correspond to $\lambda = 405/436, 546, 579$ mμ.

Table 4 gives numerical values of the scattering functions for different angles

---

[1] A solution of the approximate equation of transfer for a single given law of variation of the scattering function with height was given in [19]. An approach to the solution of (1.1) for scattering functions varying with height was given in [20], [21].

TABLE 3

$\tau^*=0,1$			$\tau^*=0,2$			$\tau^*=0,3$			$\tau^*=0,4$			$\tau^*=0,6$			$\tau^*=0,8$		
No. of scatt. func.	Γ_1/Γ_2	S_h km	No. of scatt. func.	Γ_1/Γ_2	S_h km	No. of scatt. func.	Γ_1/Γ_2	S_h km	No. of scatt. func.	Γ_1/Γ_2	S_h km	No. of scatt. func.	Γ_1/Γ_2	S_h km	No. of scatt. func.	Γ_1/Γ_2	S_h km
V	1,8	100	V	1,8	100	V	1,8	100	VI	2,5	50	VII	4,7	20	VII	4,7	20
VI	2,5	50	VI	2,5	50	VI	2,5	50	VII	4,7	20	VIII	8,4	10	VIII	8,4	10
									II	2,5	33	II	2,5	33			
									III	2,7	33	III	2,7	33			
									IV	2,3	33	IV	2,3	33			

TABLE 4

φ, °	No. of scattering function							
	I*	II	III	IV	V	VI	VII	VIII
0	$4,45\cdot10^{-3}$	$166\cdot10^{-3}$	$151\cdot10^{-3}$	$111\cdot10^{-3}$	$18,2\cdot10^{-3}$	$56\cdot10^{-3}$	$195\cdot10^{-3}$	$490\cdot10^{-3}$
10	3,80	108	100	69,5	15,2	39,5	137	365
20	3,00	37,2	30,1	29,7	9,5	24,0	86	206
30	2,40	27,7	22,2	19,4	6,6	16,0	50,5	120
40	2,00	17,7	14,6	14,6	4,9	10,8	32,0	66
50	1,73	14,8	10,6	9,79	3,8	7,8	21,0	39
60	1,49	12,5	8,14	7,74	3,1	5,9	14,3	24,0
70	1,25	10,4	7,30	6,66	2,5	4,7	10,0	15,2
80	1,12	9,01	6,40	6,03	2,2	3,9	7,6	11,0
90	1,05	7,95	5,79	5,62	1,95	3,3	6,2	8,7
100	1,07	7,42	5,51	5,34	1,90	3,2	5,6	7,6
110	1,10	6,96	5,51	5,15	1,95	3,2	5,4	6,8
120	1,15	6,74	5,51	5,15	2,1	3,3	5,3	6,4
130	1,30	7,27	5,42	5,68	2,2	3,4	5,4	6,2
140	1,45	8,17	5,42	6,41	2,4	3,6	5,6	6,3
150	1,65	8,02	5,13	5,49	2,7	4,1	5,9	6,5
160	1,90	7,87	4,99	5,11	3,1	4,7	6,1	6,7
170	2,15	6,72	4,39	4,77	3,7	5,3	6,4	6,9
180	2,30	6,20	4,14	4,74	4,3	5,8	6,5	7,0
$\sigma(0)$	0,0193	0,160	0,124	0,113	0,0386	0,0752	0,196	0,391

* The scattering function assigned to the upper layer of the atmosphere corresponds to a visibility $S_h = 200$ km and $\Gamma_1/\Gamma_2 = 1.3$.

φ which were used in the calculation. The bottom line of the table gives the corresponding values of the scattering coefficient $\sigma(0)$.

§3. Treatment of the Scattering Functions

The method of solution of Eq. (1.1) which we have used is based on the development of the experimental scattering function into a series of the form (1.2). In practice, the series should not include too large a number of terms since the number of integrodifferential equations is twice the number of terms in the series. In all our calculations we took N = 10.

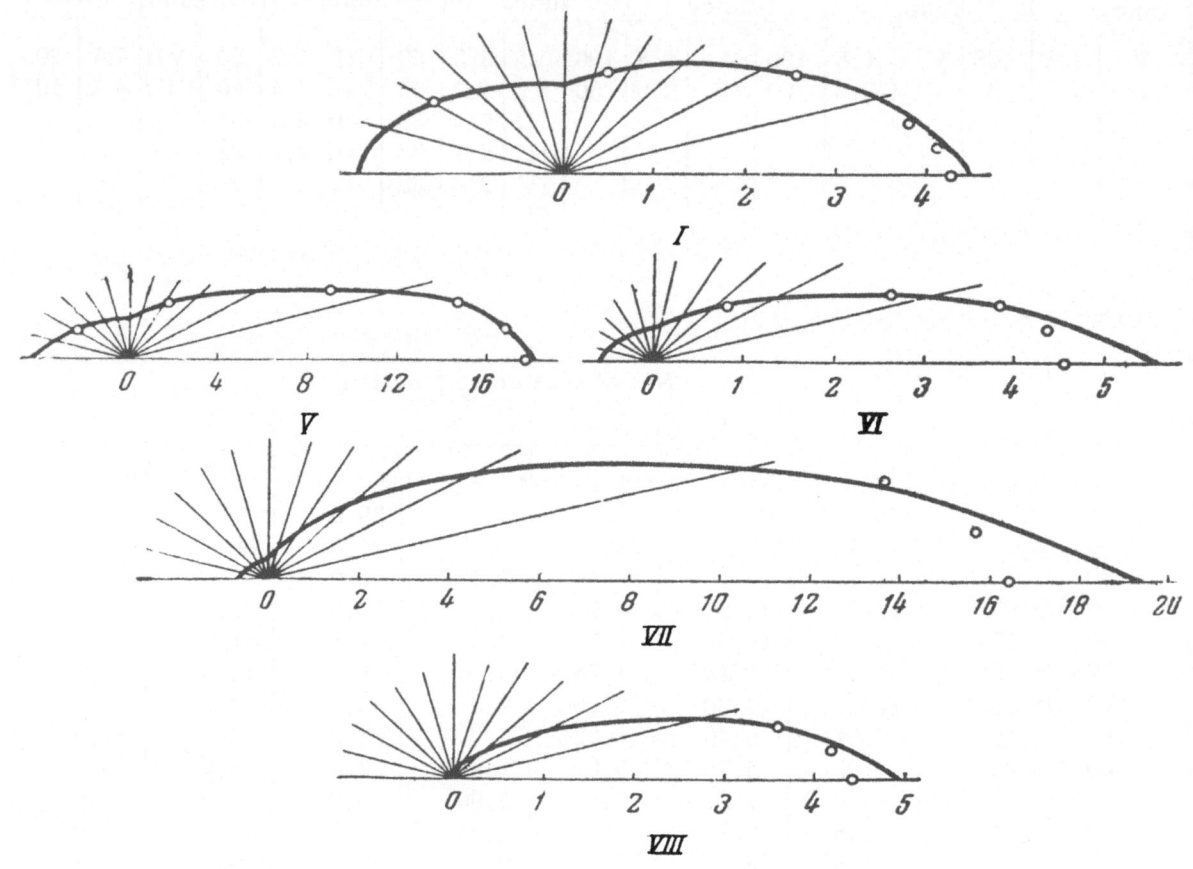

Fig. 3

Figure 3 shows the scattering functions (taken from [15]) which were used in our calculations; the points represent an approximation to them obtained by using ten terms in the series. As can be seen, even with such a large number of terms, in the range $0 \leq \varphi \leq 5°$ the true and approximate characteristics differ quite considerably. At $\varphi = 0°$ the error reaches 20% in some cases. For $5° < \varphi \leq 180°$ ten times are sufficient to represent the experimental scattering function quite well. All the same, it must be noted that the calculations described below are concerned not with the scattering functions to which they are ascribed, but to functions which are less drawn out in the small-angle region, as indicated by the points in Fig. 3. The discrepancies in the region $\varphi \leq 5°$ have only a small effect on the ratio Γ_1/Γ_2, as is shown in Table 5.

Table 6 gives the expansion coefficients for the scattering functions I,...., VIII expanded into a series (1.2); they are the Fourier coefficients determined from the formula

$$C_i = \int_0^\pi \gamma(\varphi) P_i(\cos\varphi) \sin\varphi \, d\varphi. \tag{2.5}$$

TABLE 5

No. of scattering function	I	II	III	IV	V	VI	VII	VIII
Exp. Γ_1/Γ_2	1.30	2.50	2.66	2.33	1.80	2.47	4.70	8.40
Calc. Γ_1/Γ_2	1.36	2.55	2.79	2.37	1.85	2.48	4.69	8.46

TABLE 6

No. of scatt. func.	C_1	C_2	C_3	C_4	C_5	C_6	C_7	C_8	C_9	C_{10}
I	0,3436	0,6731	0,1980	0,2245	0,1230	0,0944	0,1239	0,0217	0,0635	0,0392
II	1,0234	1,2100	1,0792	1,0461	1,0836	0,9225	0,9234	0,7741	0,6977	0,6659
III	1,1806	1,4255	1,4735	1,3759	1,3420	1,2306	1,1504	1,0677	1,0037	0,7837
IV	0,9931	1,2137	1,1163	0,9328	0,9186	0,7266	0,7567	0,7626	0,6868	0,7074
V	0,7084	1,0252	0,5804	0,5771	0,3780	0,3889	0,2559	0,2467	0,1586	0,1306
VI	1,0450	1,3224	0,9260	0,8361	0,6154	0,5448	0,3914	0,1355	0,2566	0,2312
VII	1,5786	1,7990	1,4864	1,1885	0,9225	0,7274	0,5738	0,4403	0,3358	0,2616
VIII	1,9691	2,3123	2,0879	1,7217	1,3465	1,0520	0,8247	0,6443	0,5098	0,4143

Table 6 shows that the coefficients C_i (i = 1, 2,....5) increase as the scattering curve becomes more drawn out.

§ 4.   Transition from Optical Thickness to Geometrical Height

As is known, the optical thickness of the atmosphere can be expressed in the form

$$\tau = \int_0^z \sigma(z)\, dz. \tag{2.6}$$

Formula (2.6) may be used in the transition from τ to $\underline{z}$ for each definite form of the function $\sigma(z)$. However, the experimental data of Foitzik and Zschaeck [15] which were used by us include only the values of the scattering coefficients on the earth, i.e., $\sigma(0)$, and different scattering functions. In order to get an idea as to the possible behavior of the function $\sigma(z)$ we use the experimental data of Waldram [6]. The values of the scattering coefficients $\sigma_\omega(z)$ which were obtained by Waldram were approximated exponentials using the method of least squares. In this, the following were noted. For a sufficiently transparent atmosphere in which $\Gamma_1/\Gamma_2 \lesssim 2$, the scattering coefficient $\sigma(z)$ can be well approximated to by the exponential function $\sigma(z) = $

$= \sigma_\omega e^{-kz}$, where $\sigma_0 \approx \sigma_\omega(0)$, and $k \approx \dfrac{\sigma_\omega(0)}{\tau^*}$ (Fig. 4).

With this value of τ, the curves calculated using $\sigma_\omega(z)$ (solid line) and $\sigma_1(z) = $

$= \sigma_\omega(0)\, e^{\dfrac{-\sigma_\omega(0)}{\tau^*}}\, \underline{z}$ (dotted line) are close to each other.

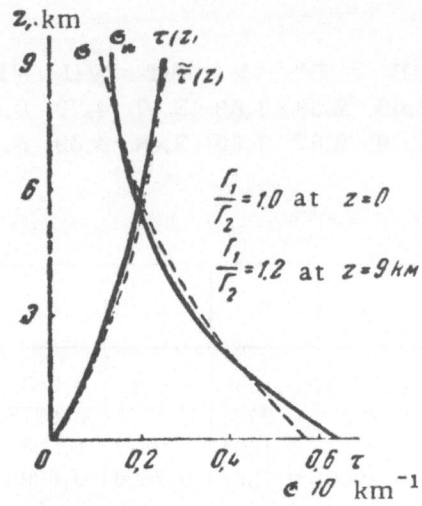

Fig. 4

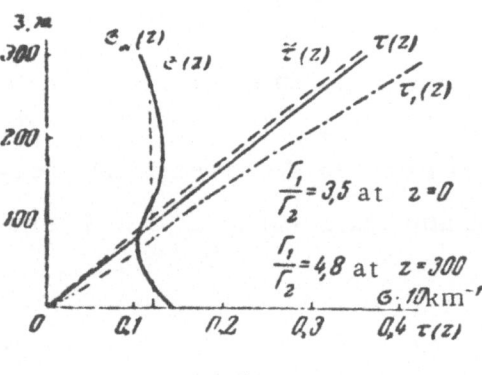

Fig. 5

For an atmosphere in which $\Gamma_1/\Gamma_2 > 2$, the exponential curve obtained by the method of least squares approaches $\sigma_\omega(z)$ (Fig. 5) only approximately, while $\tilde{\tau}(z)$ calculated from $\tilde{\sigma}(z)$ (dotted line) represents $\tau(z)$ determined from $\sigma_\omega(z)$ (solid line) rather well. If in this case also, we take $\sigma_1(z)$ as the approximating function, then in the transition from τ to $\underline{z}$ one obtains considerable deviations of the function $\tau_1(z)$ from the function $\tau(z)$ calculated from experimental data (the dot — dash curve $\tau_1(z)$ in Fig. 5).

For this reason, the transition from τ to $\underline{z}$ for the scattering curves (V–VIII) was carried out in the following way. In the case of the curves V and VI in which $\Gamma_1/\Gamma_2 \sim 2$ we assumed

$$\sigma(z) = \sigma^* \left(1 - e^{-\frac{\sigma(0)}{\sigma^*}z}\right). \quad (2.7)$$

In the case of the curves VII and VIII in which $\Gamma_1/\Gamma_2 > 2$ the transition from τ to $\underline{z}$ was carried out using formula (2.7), in which $\sigma(0)$ was taken equal to 0.124 km^{-1} for curve VII and 0.152 km^{-1} for VIII. These values were obtained from a fit to Waldram's scattering coefficients (for curves close to VII and VIII) by exponentials using the method of least squares. We used Waldram's curves for which

$$\left.\frac{\Gamma_1}{\Gamma_2}\right|_{z=0} = 3.0; \quad \sigma(0) = 0.146 \quad \text{km}^{-1}$$

and

$$\left.\frac{\Gamma_1}{\Gamma_2}\right|_{z=0} = 3.5; \quad \sigma(0) = 0.344 \quad \text{km}^{-1}$$

The latter transition from τ to $\underline{z}$ cannot be considered as very accurate, and when more complete experimental material is available, it can be improved without loss to the main data on the distribution of intensity of scattered radiation.

When τ^* is small and $\Gamma_1/\Gamma_2 \approx 2$, formula (2.7) may be used to obtain a sufficiently accurate transition τ to $\underline{z}$. Table V in the appendix gives the numerical values of the function τ (z) for the curves V-VIII and the corresponding τ^*.

CHAPTER III

RESULTS OF CALCULATIONS AND SOME DEDUCTIONS FROM THEM

§ 1. On the Convergence of Series and Successive Approximations

The most important point of the above method, from the mathematical point of view, is the rate at which the trigonometric series (1.4) and the successive approximations to its coefficients A_k (τ, θ) converge. Unfortunately it is not possible to give sufficiently accurate and general estimates of this kind for arbitrary physical parameters. We shall therefore limit ourselves only to some general discussions of the convergence of the process of successive approximations and the series (1.4), and will give graphs showing the rapidity of convergence in even most unfavorable cases.

In the case of isotropic scattering, E. S. Kuznetsov and B. V. Ovchinskii [1] have shown that the convergence of the process of successive approximations becomes less rapid as τ * increases. It is possible to show by means of relatively simple estimates that in the case of anisotropic scattering the convergence of the process of successive approximations is also less rapid as τ* increases, but, in addition, it depends on the scattering function and the direction of the ray. The convergence is less rapid for scattering curves which are more drawn out and for directions close to the horizon.

For simplicity, let us consider a single-layer atmosphere. It is not difficult to show that, for example, for the zero-order term of series of the form (1.4)

$$\left.\begin{aligned} |A_{0,\,s+1}^{(1)} - A_{0,\,s}^{(1)}| &\leqslant [\max|A_{0,\,s}^{(1)} - A_{0,\,s-1}^{(1)}| + \max|A_{0,\,s}^{(2)} - A_{0,\,s-1}^{(2)}|]\,L\,(\tau^*,\,\theta), \\ |A_{0,\,s+1}^{(2)} - A_{0,\,s}^{(2)}| &\leqslant [\max|A_{0,\,s}^{(1)} - A_{0,\,s-1}^{(1)}| + \max|A_{0,\,s}^{(2)} - A_{0,\,s-1}^{(2)}|]\,L\,(\tau^*,\,\theta), \end{aligned}\right\} \quad (3.1)$$

where $\underline{s}$ is the number of the approximation

$$L\,(\tau^*,\,\theta) = \frac{1 - e^{-\tau^*\sec\theta}}{2}\sum_{i=0}^{10}\frac{2}{2i+1}\,|\alpha_i|\cdot|C_i|\cdot|P_i\,(\cos\theta)|, \quad (3.2)$$

and

$$\alpha_i = \int_0^{\pi/2} P_i(\cos\theta')\sin\theta'd\theta''.$$

28

L (τ^*, θ) increases as τ^*, θ and C_i increase (i.e., as the scattering curve becomes more drawn out).

The above is confirmed by the graphs shown in Figs. 6 and 7, which give three successive approximations to $A_0^{(2)}$ (0,θ) (curves 1, 2, 3, respectively) for different optical thicknesses τ^* and scattering functions.

As was shown above, we have limited ourselves in our calculations to the determination of successive approximations for only the first few coefficients A_k (τ, θ) (k = 0, 1, 2, 3, depending on τ^*). The remaining A_k were taken in the first or (for large k) the zero-order approximation. It is quite clear that the error introduced by this treatment of the terms in (1.4) corresponding to large k depends on the rate at which these terms fall off.

Formulas (1.20)-(1.23) show that the rate at which the coefficients of (1.4) decrease depends on the magnitude of the residual sums of the series (1.13), (1.14). As k increases, the residual sums will decrease if the summation in (1.13), (1.14) is extended to an infinite number of terms. A treatment of the scattering functions has shown that when they are expanded into (1.2), a sufficiently large number of terms must be taken.[1] At the same time, calculations have shown that the coefficients A_k can only be calculated for the first few numbers. Only the first few terms were therefore taken accurately into account in (1.4), although in the expansion of the scattering function all 11 terms were taken into account. The greater the number of terms taken in the expansion of the scattering function, the more rapid is the decrease in A_k as k increases, and, correspondingly, the calculated value of the intensity I for some fixed number of coefficients A_k will be more accurate. As an illustration of the rate at which the terms of the series (1.4) decrease, Fig. 9 shows six successive coefficients A_k for $\tau^* = 0.4$. It is easy to see that A_5 is not more than 5% of A_0. The change of sign of the coefficients may be used to estimate the error (according to a well-known theorem) from the first of the rejected terms.[2] For small τ^* (≤ 0.4) the error does not exceed 5%, and for large τ^* ($\tau^* = 0.8$) it may reach 10-15%.

§2.   Dependence of the Intensity of the Scattered Radiation on the Altitude of the Sun, the Transparency of the Atmosphere, and the Form of the Scattering Function

The calculations which have been carried out have produced a considerable volume of material for studying regularities in the distribution of the intensity of scattered radiation with height and direction, as well as its dependence on the optical characteristics of the atmosphere. In the present work only preliminary qualitative results of a physical nature are reported. Furthermore, in all cases, except where stated otherwise, the problem is considered without taking into account reflection from the earth's surface.

---

[1] Compare with Fig. 8, where an eight-term expansion into (1.2) is shown by crosses and a ten-term expansion by circles.

[2] We must, however, note that the variability of sign in the series (1.4) does not hold for large values of k. This is due to the limited number of terms in the expansion for the scattering function.

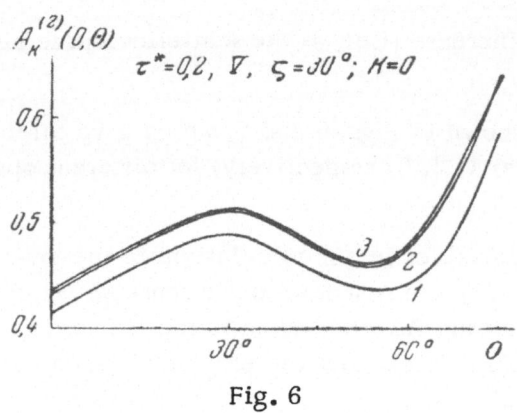

Fig. 6

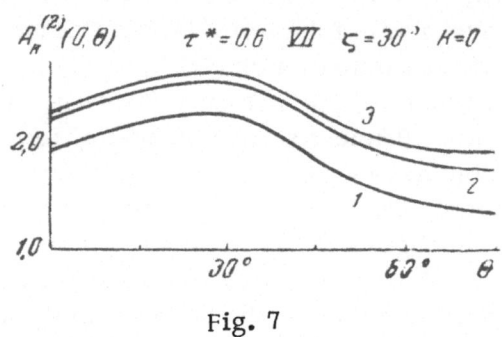

Fig. 7

A. Behavior of the Isophotes

The dependence of the intensity of radiation on direction at a given height is best represented by charts of isophotes.

Figures 10, 11, and 12 show isophotes of the brightness of the sky $I^{(2)}$ $(0, \theta, \psi)$ for small ($\tau * = 0.2$), intermediate ($\tau * = 0.4$), and large ($\tau * = 0.8$) optical thicknesses. In each figure isophotes are given for four values of the zenith distance of the sun ($\zeta =$ 30, 45, 60 and 75°) and for two scattering functions of different form.

In all these charts there is a well-defined region of the solar aureole with maximum values of brightness, and a region of minumum brightness on the side opposite to the sun. In the remaining part of the sky one observes an increase in $I^{(2)}$ $(0, \theta, \psi)$ as θ increases at $\psi =$ const, i.e., in the direction toward the horizon. With ψ increasing at $\theta =$ const, the brightness decreases.

These main results are in good agreement with observational data (cf., for example, [16] and [22]). In [22] the following results were obtained on the position of the minimum of brightness on the sky:

1. Distance between the sun and the point of minimum brightness Δ is 60-105°.

2. As the sun sets, the region of the minimum rises.

3. As the sun rises, Δ decreases, which decrease is slower for shorter wavelengths.

4. As the wavelength increases, Δ increases also.

Table 7 gives the values of Δ and θ which are the coordinates of the center of the minimum according to our calculations. These numbers should be looked upon as only preliminary rough estimates, since the calculations of θ were carried out with an interval of 15° and the error in θ may reach up to 5-10°. It is quite clear from this table that:

1. The calculated Δ lies within the limits 60-100° and in the majority of cases it lies within the interval 70-90°.

2. As the sun sets, θ decreases, i.e., the region of the minimum rises.

3. As $\tau *$ increases at large altitudes of the sun ($\zeta = 30°$), Δ increases. One of the reasons for the increase in $\tau *$ may be the decrease of λ and thus this result

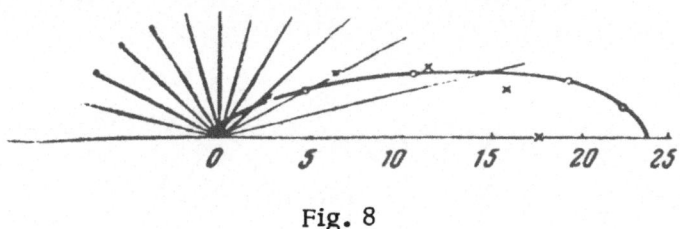

Fig. 8

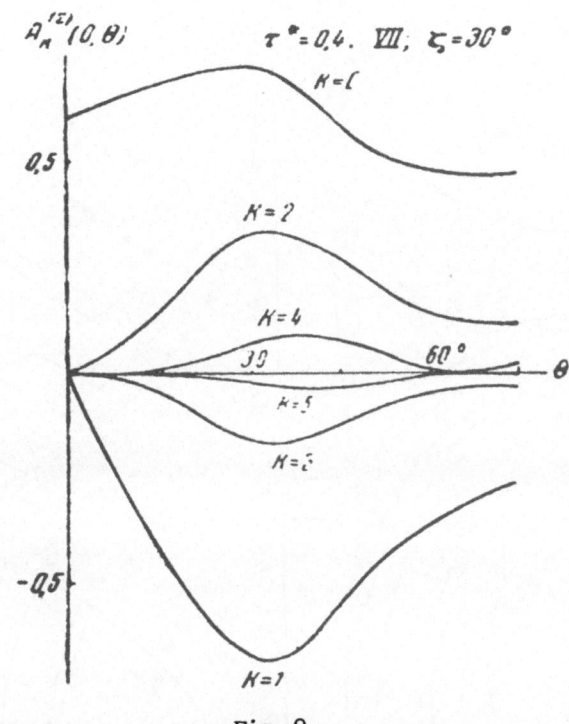

Fig. 9

TABLE 7

No. of funct.	τ^*	ζ°	30	45	60	75	No. of funct.	τ^*	ζ°	30	45	60	75
V	0,2	Δ	68	76	85	90	VII	0,4	Δ	78	87	87	91
		θ	38	31	25	15			θ	48	42	27	16
VI	0,2	Δ	68	76	86	90	VII	0,8	Δ	83	89	91	95
		θ	38	31	26	15			θ	53	44	31	20
VI	0,4	Δ	76	88	84	89	VI 11	0,8	Δ	91	88	92	91
		θ	46	43	24	14			θ	61	43	33	16

contradicts the results of I. N. Yaroslavtsev. For small altitudes of the sun Δ is almost independent of τ^*.

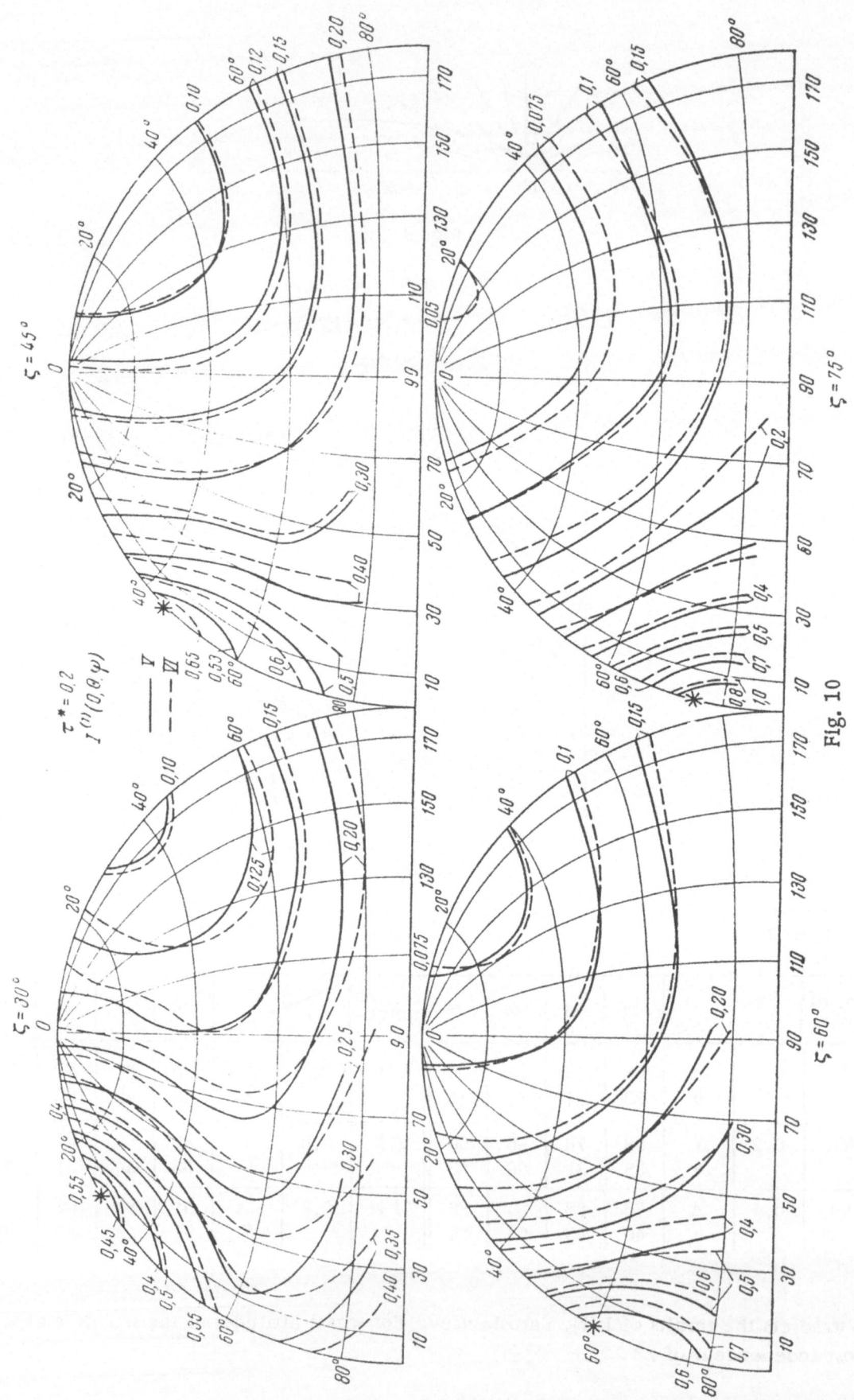

Fig. 10

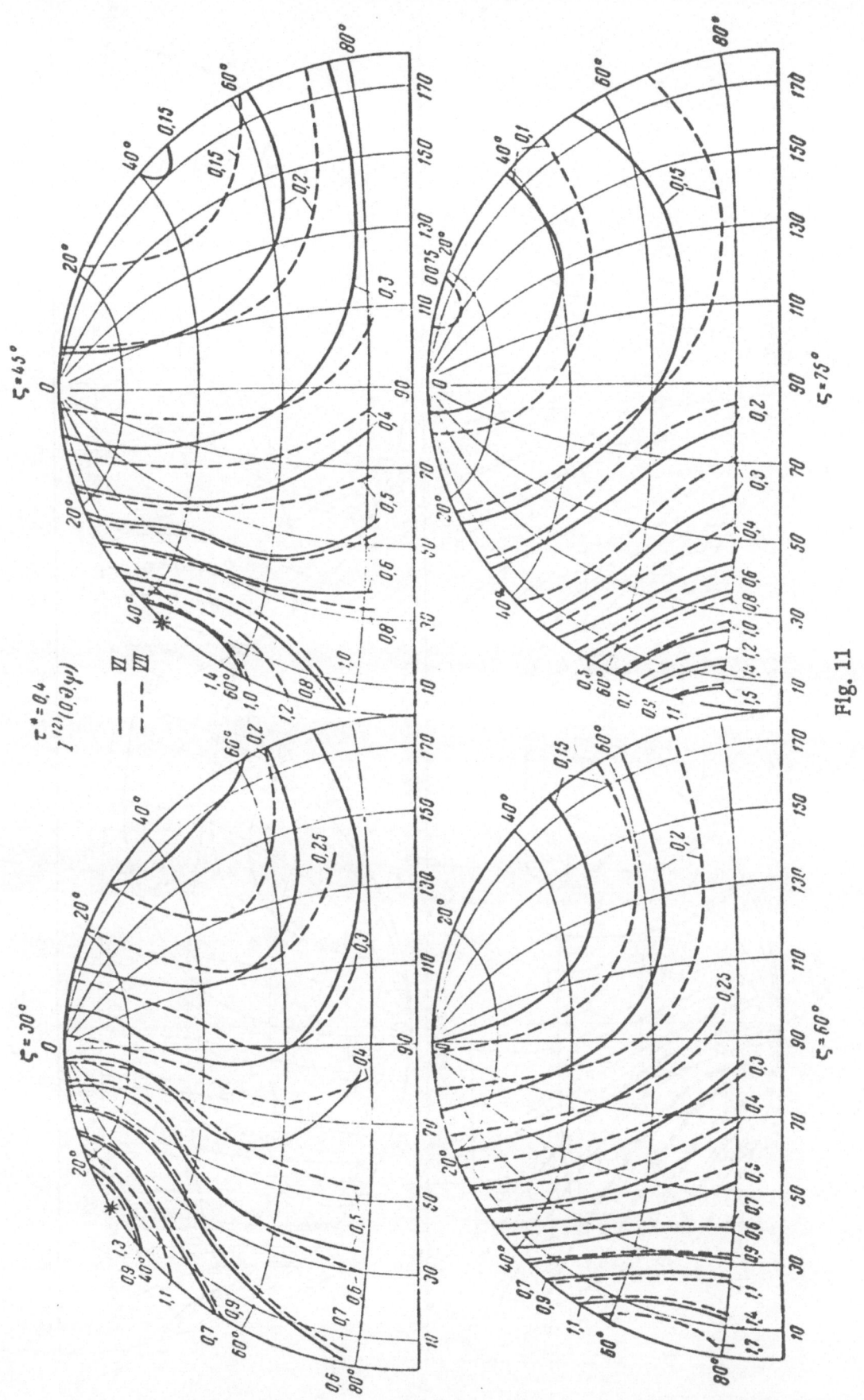

Fig. 11

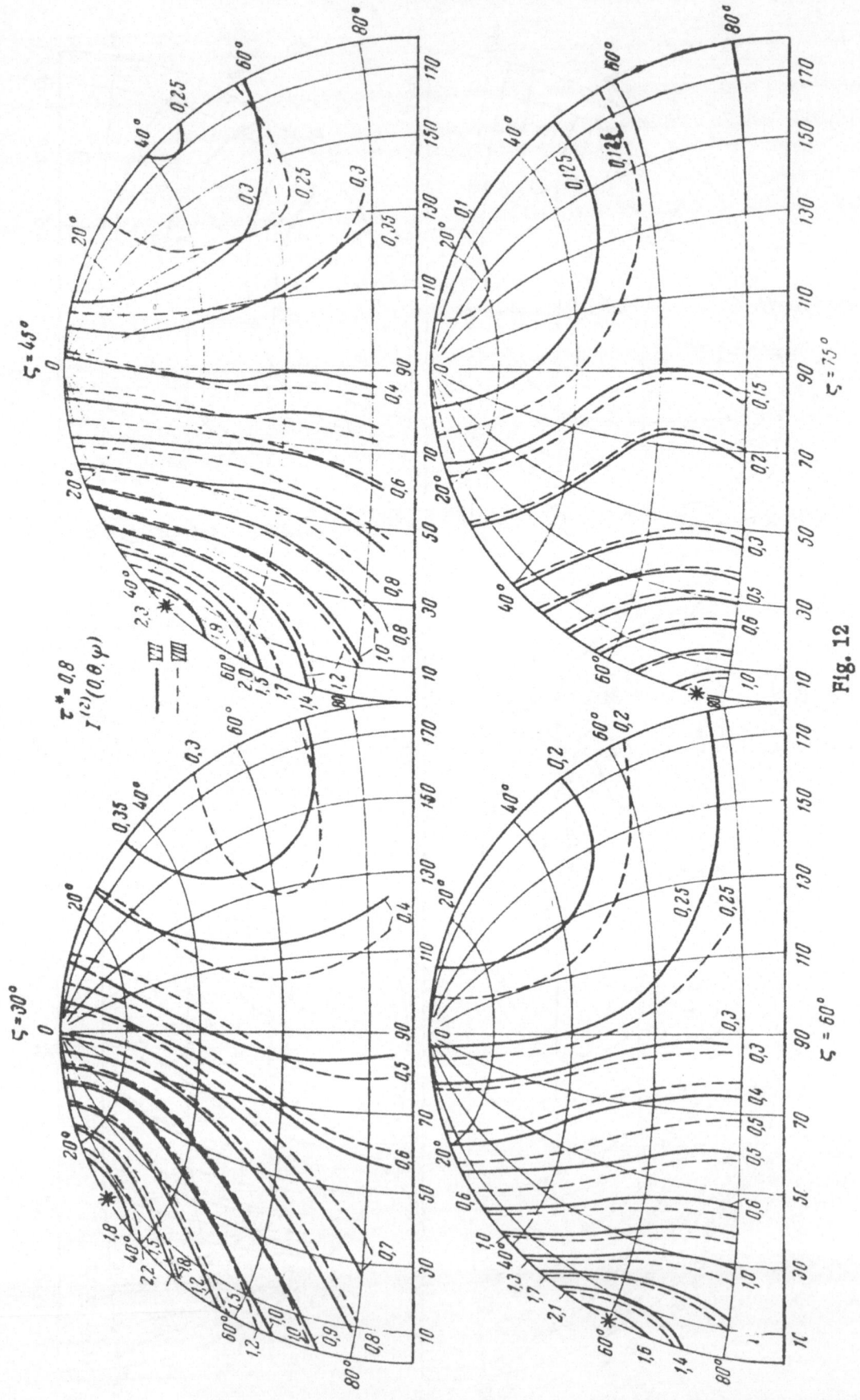

Fig. 12

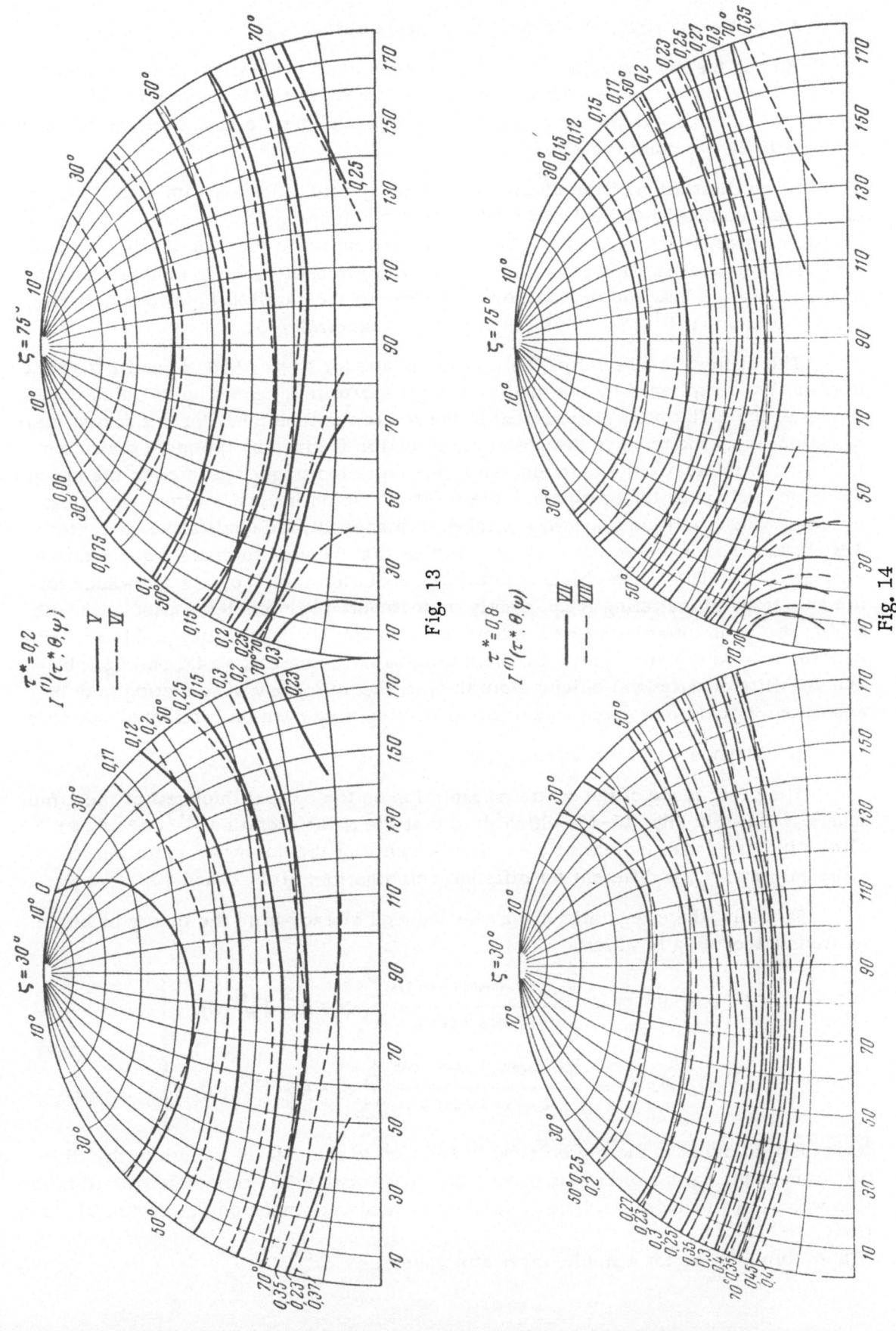

Fig. 13

Fig. 14

4. As the sun rises, Δ almost always decreases.

Among the main regularities which may be noted in Figs. 10-12 is the change in the form of the isophotes with changes in the zenith distance of the sun. The larger the zenith distance, the smoother are the isophotes and the closer are they to the co-ordinate lines θ = const (parallels).

As an illustration of the distribution of the upward radiation with direction, charts were computed of isophotes of the radiation escaping at the upper boundary of the atmosphere, $I^{(1)}(\tau^*, \theta, \psi)$. These charts are shown in Figs. 13 and 14. These isophotes are much smoother than the $I^{(2)}(0, \theta, \psi)$ isophotes, the azimuthal variation of the intensity is less and the isophotes are close to the parallels both for small and large values of ζ. $I^{(1)}(\tau^*, \theta, \psi)$ increases with increasing θ.

The different character of the isophote charts for $I^{(1)}(\tau^*, \theta, \psi)$ and $I^{(2)}(\tau^*, \theta, \psi)$ may easily be explained by the theory of single scattering, according to which the intensity of the radiation is proportional to the scattering function. For this reason solar radiation singly scattered in the downward direction is given by the more drawn out front part of the scattering function, while the radiation singly scattered in the upward direction is given by the rear part of the function, which is usually almost spherical in form. When multiple scattering is taken into account, the qualitative character of this phenomenon is preserved. Our calculations may be used to make a quantitative estimate of it. The above charts lead to the conclusion that accurate allowance for the anisotropy of scattering is apparently most important in problems associated with studies of the downward scattered radiation, for example, in visibility problems. Whenever the upward radiation plays the main role, as in the case of aerial photography and visibility of terrestrial objects from the air, the anisotropy of scattering may be accounted for only in a rough way or the scattering may even be considered isotropic.[1]

B. Dependence of the Intensity on τ^* and ζ

The dependence of the scattered radiation on the optical thickness of the atmosphere is shown in Figs. 15-18, which show that the rising radiation $I^{(1)}(\tau^*, \theta, \psi)$ always increases with increasing τ^*. The behavior of the downward radiation for large values of τ^* is different for different combinations of ζ, θ and ψ.

Basically, these regularities may be deduced already from the theory of single scattering according to which[2]

$$\left.
\begin{aligned}
I^{(1)}(\tau^*,\ 0,\ \psi) &= \frac{1 - e^{-\tau^*(\sec\theta + \sec\zeta)}}{1 + \cos\theta\,\sec\zeta} \sum_{k=0}^{N} \cos k\psi \cdot F_k^{(1)}(0), \\
I^{(2)}(0,\ 0,\ \psi) &= \frac{e^{-\tau^*\sec\zeta} - e^{-\tau^*\sec\theta}}{1 - \cos\theta\,\sec\zeta} \sum_{k=0}^{N} \cos k\psi \cdot F_k^{(2)}(0).
\end{aligned}
\right\} \tag{3.3}$$

---

[1] One must note that in these problems, in addition to the upward radiation, the illumination of the earth's surface is important. It is equal to the flux of downward radiation which, as will be shown below, does not depend very much on the degree of anisotropy.

[2] These formulas are for a single-layer atmosphere.

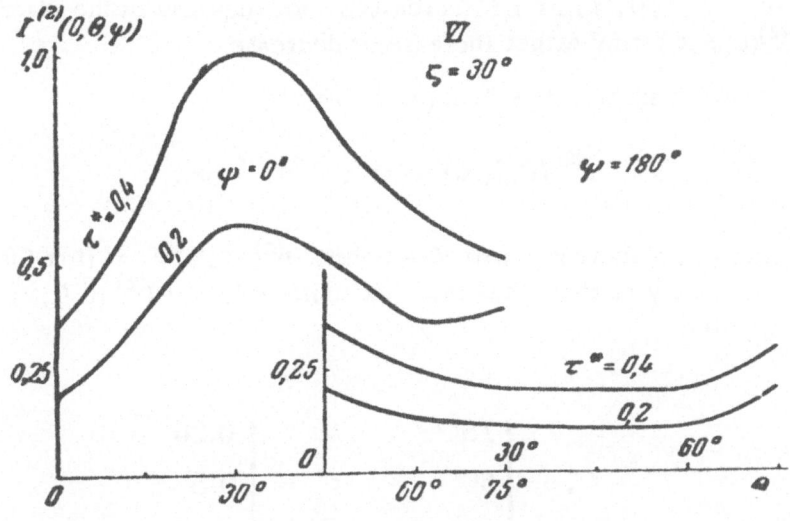

Fig. 15

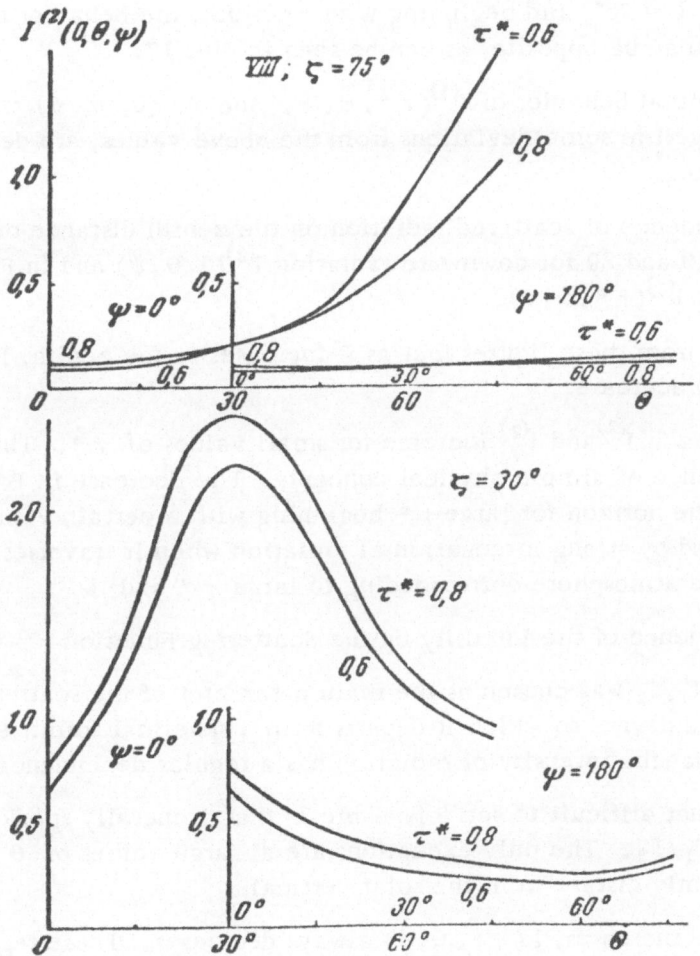

Fig. 16

The behavior of $I^{(1)}(\tau^*, \theta, \psi)$ as τ^* increases is not required in the discussion; according to (3.3), $I^{(2)}(0, \theta, \psi)$ may either increase or decrease.

Let us put, for example, $\theta = \zeta$ in (3.3). Hence

$$I^{(2)}(0, \zeta, \psi) = \tau^* \, e^{-\tau^* \sec \zeta} \sec \zeta,$$

and it follows that as τ^* increases with $\zeta = $ const., $I^{(2)}(0, \zeta, \psi)$ at first increases and then decreases. It is easy to show that the maximum value of $I^{(2)}(0, \zeta, \psi)$ is reached at the following points.

$$\zeta = \begin{cases} 75° \\ 60° \\ 30° \end{cases} \qquad \tau^* = \begin{cases} 0.26 \\ 0.5 \\ 0.87. \end{cases}$$

Thus for $\zeta = 30°$, $I^{(2)}(0, \zeta, \psi)$ will increase for all the values of τ^* which we considered. For $\zeta = 75°$, and beginning with $\tau^* = 0.3$, the behavior of $I^{(2)}(0, \theta, \psi)$ as τ^* increases may be opposite, as can be seen in Fig. 17.

The azimuthal behavior of $I^{(1)}(\tau^*, \theta, \psi)$ and $I^{(2)}(0, \theta, \psi)$ with a varying τ^*, as well as some deviations from the above values, are determined by multiple scattering.

The dependence of scattered radiation on the zenith distance of the sun is illustrated in Figs. 19 and 20 for downward radiation $I^{(2)}(0, \theta, \psi)$ and in Figs. 21 and 22 for upward radiation $I^{(1)}(\tau^*, \theta, \psi)$.

It is clear from these figures that as ζ increases in the zenith, $I^{(1)}(\tau^*, \theta, \psi)$ and $I^{(2)}(0, \theta, \psi)$ decrease.

At the horizon $I^{(1)}$ and $I^{(2)}$ increase for small values of τ^*. This result is a natural consequence of simple physical concepts. The decrease in $I^{(1)}(\tau^*, \theta, \psi)$ and $I^{(2)}(0, \theta, \psi)$ at the horizon for large τ^* beginning with a certain value of ζ, is evidently determined by strong attenuation of radiation when it traverses considerable thicknesses of the atmosphere corresponding to large τ^* and ζ.

C. Dependence of the Intensity on the Scattering Function

The ratio Γ_1/Γ_2 was chosen as the main parameter of the scattering function and represents the degree to which it departs from a spherical form. Results of calculations show that the intensity of radiation has a regular dependence of Γ_1/Γ_2.

Thus it is not difficult to see (cf. Table 1) that, generally speaking, $I^{(2)}(0, \theta, \psi)$ increases with Γ_1/Γ_2. The only exceptions are at large values of θ ($> 60°$) in verticals sufficiently distant from the solar vertical.

As $\Gamma_1 \Gamma_2$ increases, $I(\tau^*, \theta, \psi)$ always decreases. These regularities are preserved also at all inner points of the atmosphere when $0 \le \tau \le \tau^*$. It is clear from physical arguments that this is precisely the way in which the intensity should vary as the scattering function becomes more anisotropic.

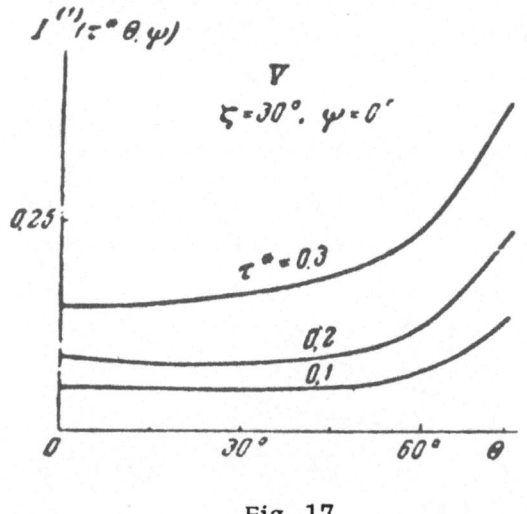

Fig. 17

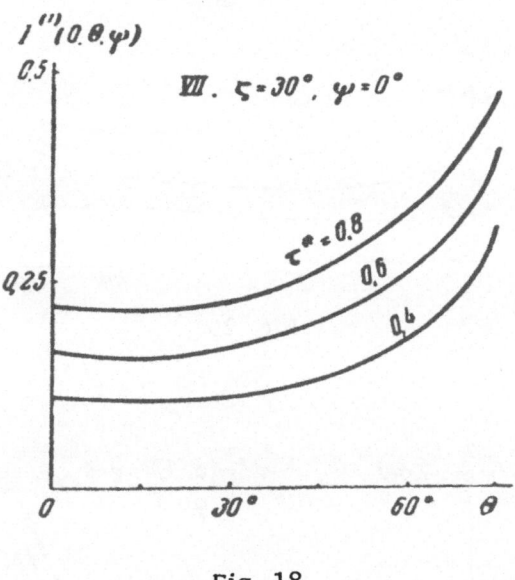

Fig. 18

Evidently, the ratio Γ_1/Γ_2 may taken as a certain characteristic of the scattering function, convenient in the elucidation of regularities in the behavior of the intensity with changes in the form of this function. At the same time it is understandable that this characteristic, being only an integral characteristic, does not have the flexibility which is necessary in the description of finer effects. For example, it cannot be used in the explanation of the regularities in the changes of downward radiation in the solar vertical, which is very sensitive to changes in the scattering function.

Figures 23 and 24 show the brightness of the sky in the solar vertical for $\tau^* = 0.4$ and 0.6 corresponding to different scattering functions. The curves are marked by the number of the scattering function and the value of Γ_1/Γ_2. It is clear that function VI gives a lower brightness than functions IV, II and III for roughly the same values of Γ_1/Γ_2, while function VII ($\Gamma_1/\Gamma_2 = 4.7$) corresponds to lower brightness than III ($\Gamma_1/\Gamma_2 = 2.8$), which contradicts the regularities established above.

In order to understand this it is necessary to introduce additional characteristics of the scattering function. $\gamma(0)$ and $\gamma(\pi)$ may be taken as such and rerespresent scattering in the direction of the incident ray and in the direction opposite to it, respectively.

Table 8 shows that function III gives a maximum value of $I^{(2)}(0,\theta,\psi)$ since to it corresponds maximum $\gamma(0)$. But this is insufficient: functions II and VII are associated with equal forward scattering (10.3-10.4), but the values of $I^{(2)}(0,\zeta,0)$ which correspond to them are different. This may be explained by the values of Γ_1/Γ_2 for these functions: the much greater value of Γ_1/Γ_2 for function VII leads to the shift of the whole curve of $I^{(2)}(0,\theta,0)$ toward larger values. Apparently, $\gamma(0)$ is responsible for the magnitude of $I^{(2)}(0,\theta,0)$, while Γ_1/Γ_2 determines the level of the whole curve of $I^{(2)}(0,\theta,0)$. In a similar way the quantities $\gamma(\pi)$ and Γ_1/Γ_2 explain the behavior of the curve in Fig. 25 which shows $I^{(1)}(\tau^*,\theta,\psi)$ as a function θ.

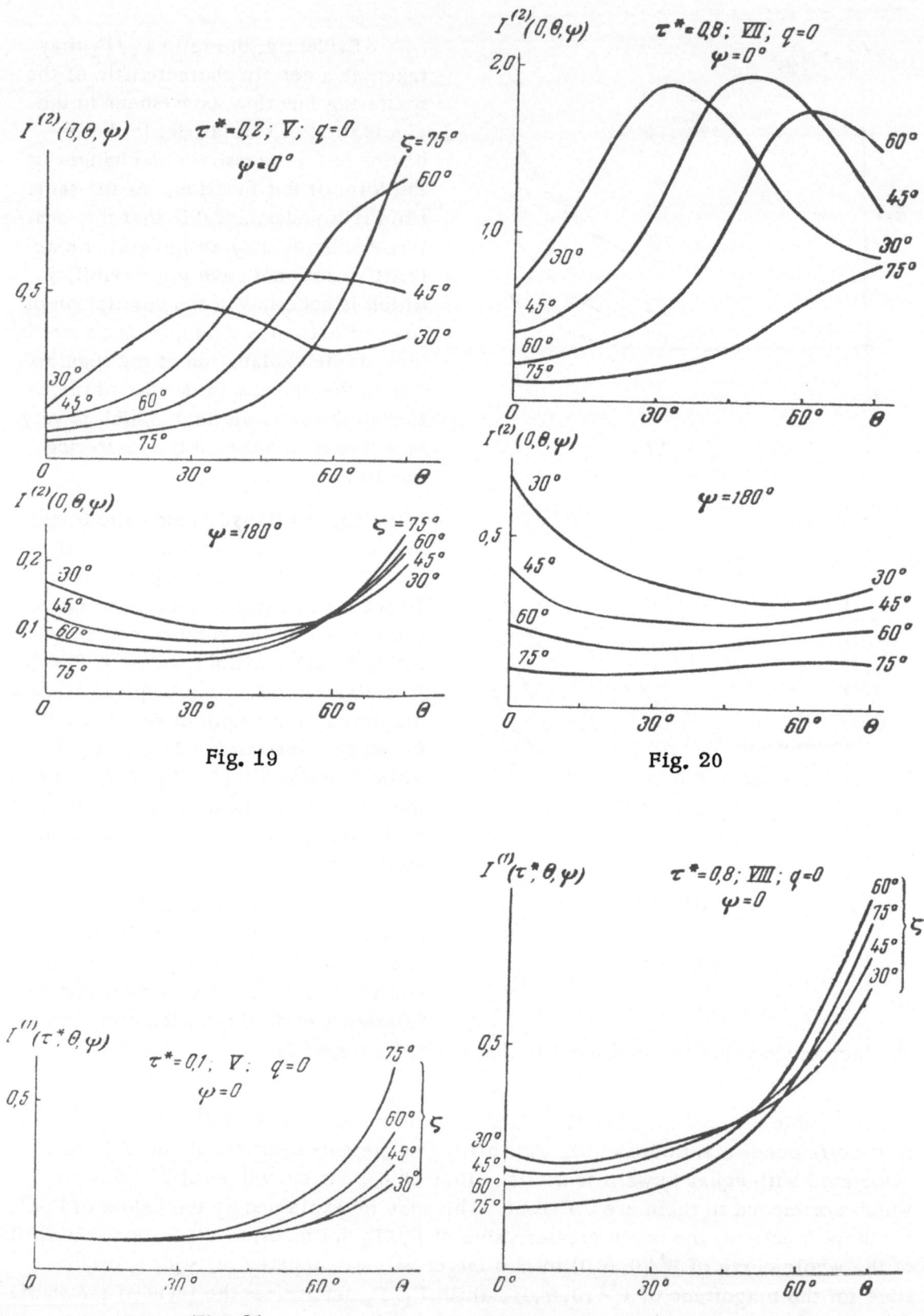

Fig. 19

Fig. 20

Fig. 21

Fig. 22

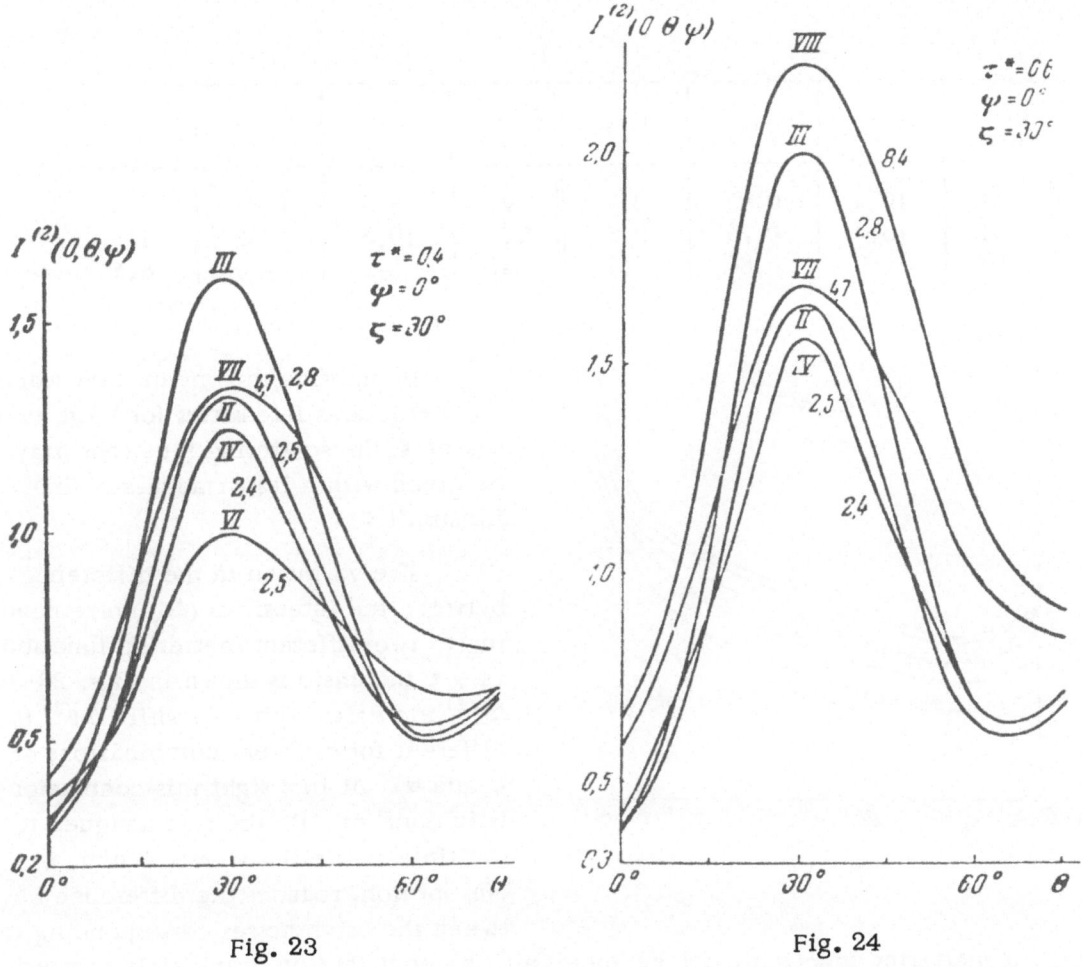

Fig. 23   Fig. 24

Quantitative changes in the intensity of radiation due to a change in the scattering function depend on $\tau*$ and ζ.

The charts of isophotes shown in Figs. 10-14 give curves corresponding to two different scattering functions. The charts clearly show that as ζ increases, pairs of isophotes corresponding to equal values of intensity but different scattering functions approach each other, and, as was already stated above, tend to the isophotes for spherical scattering function.

More accurate quantitative information on the difference between brightnesses corresponding to the two scattering functions, as the zenith distance of the sun varies, is given by Figs. 26 and 27, which show that, with the exception of large values of θ, the brightnesses of the downward radiation corresponding to the different scattering functions are very nearly the same for sufficiently large ζ. This very interesting fact, namely, the decrease in the brightness difference is evidently connected with the traversal of a large thickness of the atmosphere and, consequently with an increase in the contribution due to multiple scattering for large values of θ. Multiple scattering leads to an "averaging" over the separate acts of scattering and removes the difference between the scattering functions.

TABLE 8

No. of funct.	$\gamma(0)$	$\gamma(\pi)$	Γ_1/Γ_2	No. of funct.	$\gamma(0)$	$\gamma(\pi)$	Γ_1/Γ_2
II	10,4	0,82	2,5	VI	8,8	0,83	2,5
III	13,0	0,75	2,8	VII	10,3	0,52	4,7
IV	9,8	0,86	2,4	VIII	—	—	8,4

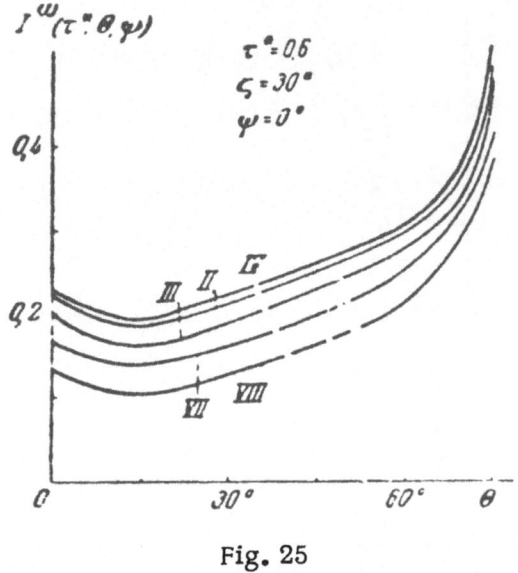

Fig. 25

In practice, this means that when one calculates intensities for large values of ζ, the scattering functions may be given with a lesser accuracy than for small ζ.

The variation in the difference between the intensities (ΔI) corresponding to two different scattering functions as τ^* increases is shown in Figs. 28-30. $\Delta I^{(1)}$ increases with τ^*, while $\Delta I^{(2)}$ is different for different combinations of ζ and θ. At first sight this conclusion is in conflict with the role assigned to multiple scattering above, which, in our opinion, reduces the difference between the brightnesses corresponding to different scattering functions. As τ^* increases, the contribution of multiple scattering grows unconditionally. The problem is resolved if one considers the formulas giving the intensity for single scattering. For simplicity we shall consider the case of a single-layer atmosphere in which the quantities $\Delta I^{(i)}$ (i = 1, 2) are described by formulas analogous to (3.3) (F is replaced by ΔF).

As can be seen, $\Delta I^{(1)}$ increases sufficiently rapidly with τ^*, $\Delta I^{(2)}$ as well as $I^{(2)}$ (cf. page 26) may in a certain interval of τ^* increase. The values of τ^* at which $\Delta I^{(2)}$ reaches its maximum depend on ζ and θ.

Evidently, the opposite effect of multiple scattering is weaker and may only reduce, but cannot completely distort, these regularities.

§ 3. Reflection of Light from the Earth's Surface

The effect of reflecting properties of the earth's surface on the magnitudes of intensity can be seen in Tables 9 and 10, in which $I^{(1)}(\tau^*, \theta, \psi)$ and $I^{(2)}(0, \theta, \psi)$ are compared for two different scattering functions and different albedo (relative differences between them are also given).

We can see that as q increases, $I^{(1)}(0, \theta, \psi)$ rapidly increases, but the rate at which $I^{(2)}(0, \theta, \psi)$ increases is much slower.[1]

[1] For large values of q the rate of change of $I^{(1)}$ and $I^{(2)}$ is roughly the same.

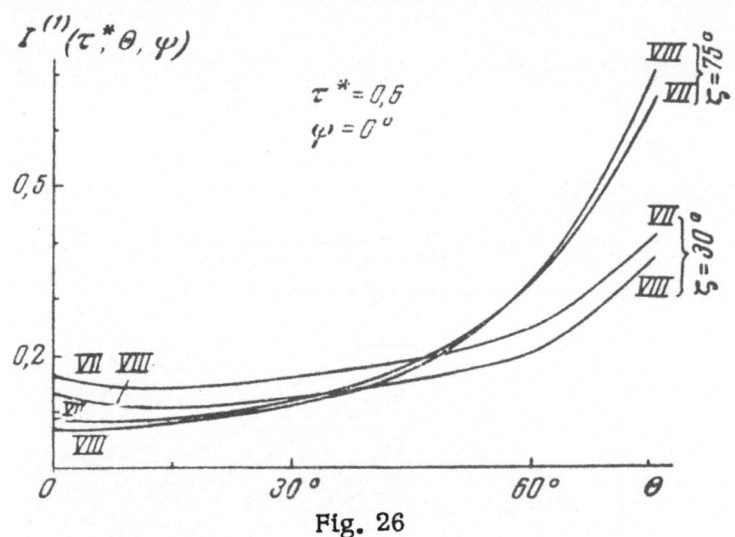

Fig. 26

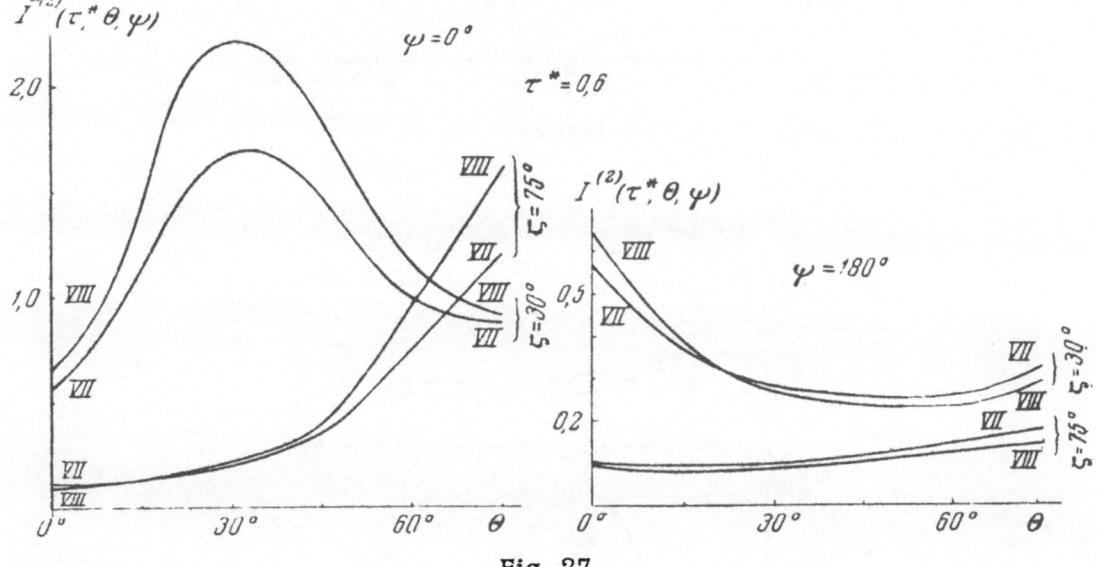

Fig. 27

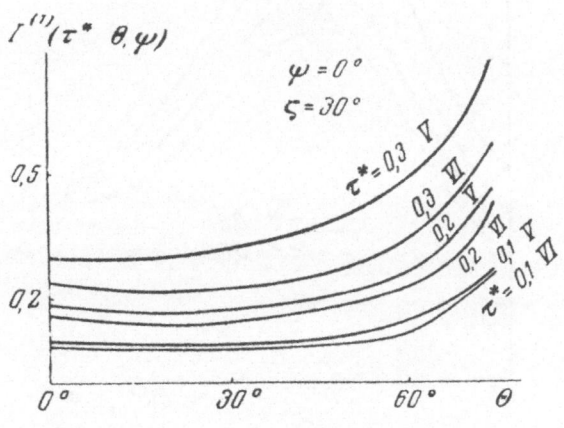

Fig. 28

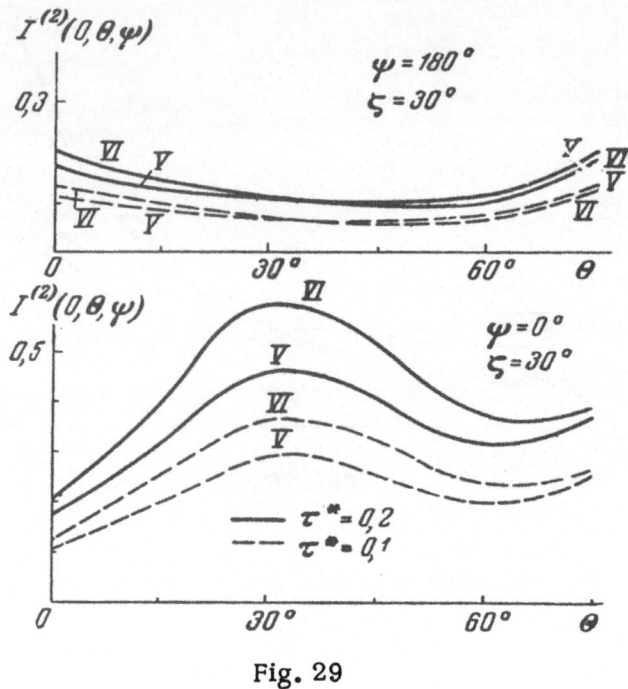

Fig. 29

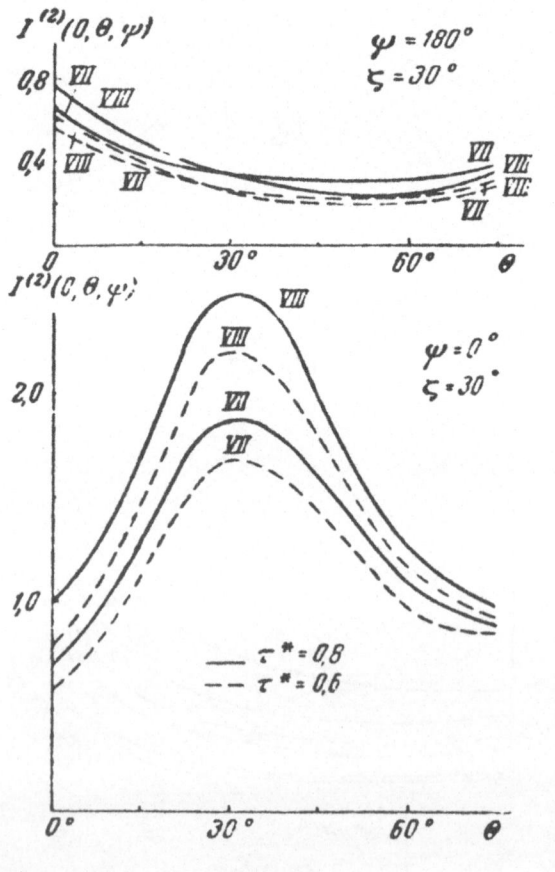

Fig. 30

This in itself is trivial but with it is associated another much more interesting fact: as the albedo increases the magnitude of the intensities $I^{(1)}$ which correspond to the different scattering functions approach each other and the effect of anisotropy is smaller.

It follows that calculations which take into account the albedo of the underlying surface confirm to an even greater degree the conclusion reached above, namely, that the role of anisotropy of scattering in intensity calculations in the case of upward-scattered radiation is relatively unimportant.

Table 10 shows that the quantities $\Delta I^{(2)}$ for downward-scattered radiation vary with increasing $\underline{q}$ to a lesser extent, although they decrease somewhat, so that the effect of anisotropy of scattering in the case of downward radiation is not considerably reduced. The above role of the reflecting properties of the underlying surface, which manifests itself in a reduction in the effective anisotropy, is connected with the assumption that Lambert's law holds.

The effect of the albedo on the distribution of $I^{(1)}(\tau,\theta,\psi)$ and $I^{(2)}(\tau,\theta,\psi)$ with height is shown in Fig. 31, from which it is clear that $I^{(1)}$ is almost independent of $\underline{q}$ at all heights right up to the upper boundary of the atmosphere.

As $\underline{q}$ increases, the difference between the quantities $I^{(1)}$ corresponding to the two scattering functions decreases.

Thus the conclusion about the relatively unimportant role of anisotropy in the calculation of $I^{(1)}(\tau^*,\theta,\psi)$ must therefore be supplemented by the result that it is essential to take accurately into account the albedo of the earth's surface.

Figure 31 shows that the quantity $I^{(2)}(\tau,\theta,\psi)$ also experiences the effect of the albedo up to considerable heights, but quantitative corrections in this case are small.

§ 4. Flux of the Scattered Radiation

The calculations which have been carried out by us may be used, within the limits of our formulation of the problem, to determine the upward and downward fluxes of scattered radiation, their distribution with height, and regularities in their behavior when changes take place in the optical characteristics of the atmosphere.

The fluxes are calculated from the formula

$$F_i(\tau) = \int_0^{2\pi} \int_0^{\pi/2} I^{(i)}(\tau,\theta,\psi) \cos\theta\, d\theta\, d\psi \qquad (3.4)$$

and for a reflecting surface whose albedo $\underline{q}$ differs from zero (3, 7) we have

$$F_{iq}(\tau) = F_i(\tau) + \pi C_0 \int_0^{\pi/2} B_0^{(i)}(\tau,\theta)\, \sin 2\theta\, d\theta. \qquad (3.5)$$

TABLE 9

$I^{(1)}(\tau^*, \theta, \psi)$

$\tau^* = 0,8; \ \psi = 0°$ $\tau^* = 0,8; \ \psi = 180°$

q	No. of funct.	ζ = 30° θ = 0°	δ, %	θ = 75°	δ, %	ζ = 75° θ = 0°	δ, %	θ = 75°	δ, %	ζ = 30° θ = 0°	δ, %	θ = 75°	δ, %	ζ = 75° θ = 0°	δ, %	θ = 75°	δ, %
0	VII	0,2227	22,0	0,4952	8,3	0,1072	14,4	0,6452	5,9	0,2227	22,0	0,4468	15,0	0,1072	14,4	0,3610	3,1
	VIII	0,1786		0,4557		0,0928		0,6846		0,1786		0,3845		0,0928		0,3501	
0,2	VII	0,4530	5,1	0,6333	3,6	0,1420	8,7	0,6661	6,1	0,4530	5,1	0,5849	8,0	0,1420	8,7	0,3819	2,3
	VIII	0,4303		0,6112		0,1302		0,7077		0,4303		0,5400		0,1302		0,3732	
0,6	VII	1,0055	1,6	0,9647	1,2	0,2254	3,4	0,7161	6,2	1,0055	1,6	0,9163	1,2	0,2254	3,4	0,4319	1,0
	VIII	1,0213		0,9762		0,2179		0,7619		1,0218		0,9050		0,2179		0,4274	

TABLE 10

$I^{(2)}(0, \theta, \psi)$

$\tau^* = 0,8; \ \psi = 0°$ $\tau^* = 0,8; \ \psi = 180°$

q	No. of funct.	ζ = 30° θ = 0°	δ, %	θ = 75°	δ, %	ζ = 75° θ = 0°	δ, %	θ = 75°	δ, %	ζ = 30° θ = 0°	δ, %	θ = 75°	δ, %	ζ = 75° θ = 0°	δ, %	θ = 75°	δ, %
0	VII	0,6854	13,1	0,8730	10,4	0,1155	5,4	0,8117	28,0	0,6854	13,1	0,3606	6,3	0,1155	5,4	0,1370	6,7
	VIII	0,7822		0,9692		0,1094		1,0768		0,7822		0,3387		0,1094		0,1280	
0,2	VII	0,7295	12,2	1,0097	8,8	0,1222	5,4	0,8323	27,4	0,7295	12,2	0,4973	5,3	0,1222	5,4	0,1576	6,4
	VIII	0,8246		1,1022		0,1157		1,0966		0,8246		0,4717		0,1157		0,1478	
0,6	VII	0,8354	10,1	1,3378	5,6	0,1381	5,7	0,8819	25,7	0,8354	10,1	0,8254	5,2	0,1381	5,7	0,2072	6,5
	VIII	0,9243		1,4144		0,1305		1,1429		0,9243		0,7839		0,1305		0,1941	

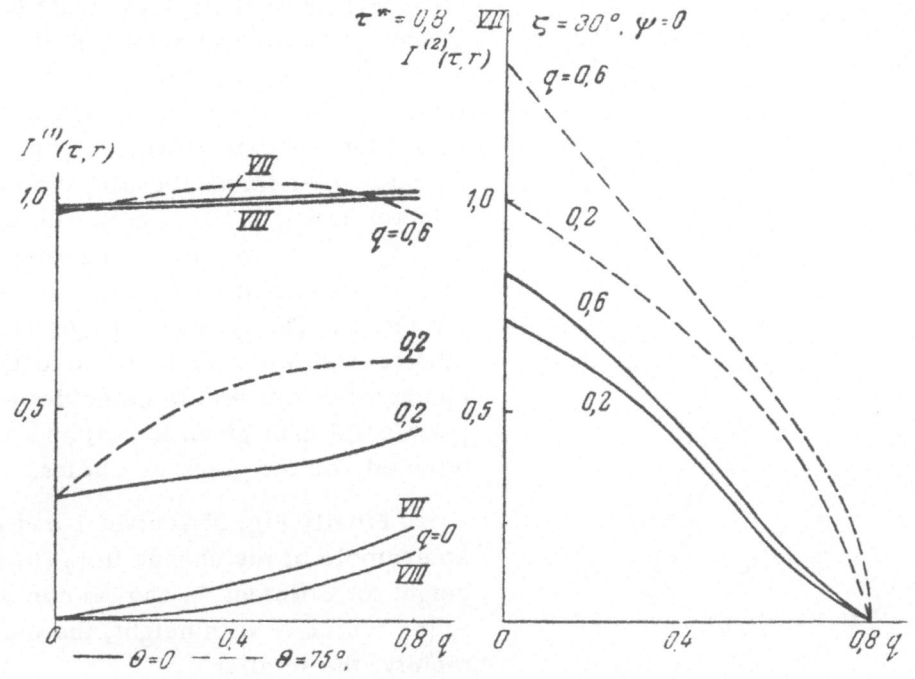

Fig. 31

A short review of the experimental studies of fluxes of scattered radiation, mainly the downward flux which reaches the earth's surface F_2 (0), is given in the monograph by K. Ya. Kondrat'ev [23]. This author shows, using observational data, how the quantity F_2 (0) changes as a function of the altitude of the sun, transparency of the atmosphere, and the albedo of the underlying surface.

We have verified all these functional dependencies and have shown that they are all confirmed by the theory. Figure 32 (curves 1, 2, and 3) shows that F_2 (0) quite rapidly decreases as the zenith distance of the sun increases, and this reduction is accelerated by an increase in the turbidity of the atmosphere (quantities τ * on each curve). Curves 4 and 5 (dotted) in Fig. 32 represent experimental values of F_2 (0) taken from [23] and correspond to the same value of τ * as curves 1 and 2. Curves 6 and 7 repeat curves 1 and 2 in relative units, which were chosen so that the left-hand ends coincide with the corresponding points of curves 4 and 5. A comparison of the pairs of curves 6, 7 and 4, 5 shows that the theory not only confirms the main regularities in changes of F_2 (0) with ζ but also gives very closely the experimental qualitative behavior of the curves. Curve 8 is given for comparison and it represents F_2 (0) for a spherical scattering function in the same relative units. The variation of F_2 (0) in this case takes place at a considerably slower rate. The quantitative divergence (almost by a factor of 2) between the theoretical and experimental values (curves 1, 2 and 4, 5, respectively) may possibly be explained by the fact that the former were obtained by multiplying the relative calculated data by the magnitude of the integral solar constant $\pi S = 1.94$ cal/cm$^2 \cdot$ min, i.e., it was assumed that the coefficient of scattering was independent of wavelength. Furthermore, the theory does not take into account absorption which reduces the measured integral flux of solar radiation.

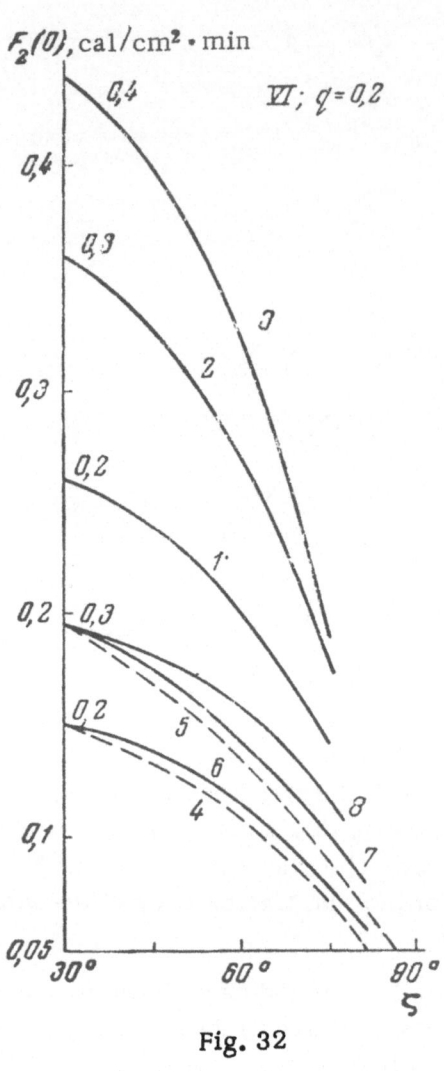

$F_2(0)$, cal/cm² · min

$\underline{VI}$; $q = 0.2$

Fig. 32

Figure 33 shows that as the turbidity of the atmosphere increases, $F_2(0)$ (curves 1, 2) increases, and this increase is more rapid, the lower the zenith distance of the sun. The slow rate at which $F_2(0)$ increases for large ζ is connected with the effect of a factor acting in the opposite direction, namely, the increasing attenuation of the radiation along the path which is equal to $\tau^* \sec \zeta$. The growth of $F_2(0)$ as $\underline{q}$ increases is shown in Fig. 34 (curve 1). The character of this growth coincides with experimental data given in [23] and is represented on the graph by circles.

Finally Fig. 35 (curves 1 and 2) gives an example of the change in $F_2(0)$ with height for different ζ, and, as can be seen, $F_2(0)$ decreases with height, the more rapidly, the smaller ζ.

Together with the curves of $F_2(0)$, each of these figures, as well as Fig. 36, shows curves of $F_1(\tau^*)$, which is equal to the flux of the scattered radiation escaping at the upper boundary of the atmosphere (3 and 4 in Fig. 33; 2 in Fig. 34; 3 and 4 in Fig. 35). $F_1(\tau^*)$, as well as $F_2(0)$, rapidly falls off as ζ increases, and the character of this reduction is not very dependent on the turbidity of the atmosphere, in accordance with the relatively slow variation in $F_1(\tau^*)$ with τ^*. $F_1(\tau^*)$ varies even more slowly with height.

The distinctive property of $F_1(\tau^*)$ is its unusually rapid increase with $\underline{q}$ (cf. curve 2 in Fig. 34). As a result of this, already at $q = 0.2$ (cf. Fig. 33) $F_1(\tau^*)$ turns out to be greater than $F_2(0)$, particularly for small τ^* and ζ, while for $q = 0$ the opposite relation holds.

Let us now consider the dependence of the flux of scattered radiation on the form of the scattering function. Tables 11 and 12 give the values of $F_1(\tau)$ and $F_2(0)$ corresponding to different albedos and relative differences δF_i (%) of the values of fluxes for two scattering functions in almost identical conditions. In addition, $F_2(0)$ is given for isotropic scattering. The corresponding quantities are taken from [1]. These tables show that $F_1(\tau^*)$ falls off with increasing Γ_1/Γ_2, while $F_2(0)$ increases, and the former varies more rapidly than the latter.

It must, however, be pointed out that we are considing here only relative differences ΔF_i. The absolute differences between the fluxes corresponding to two different scattering functions are always larger in the case of downward flux.

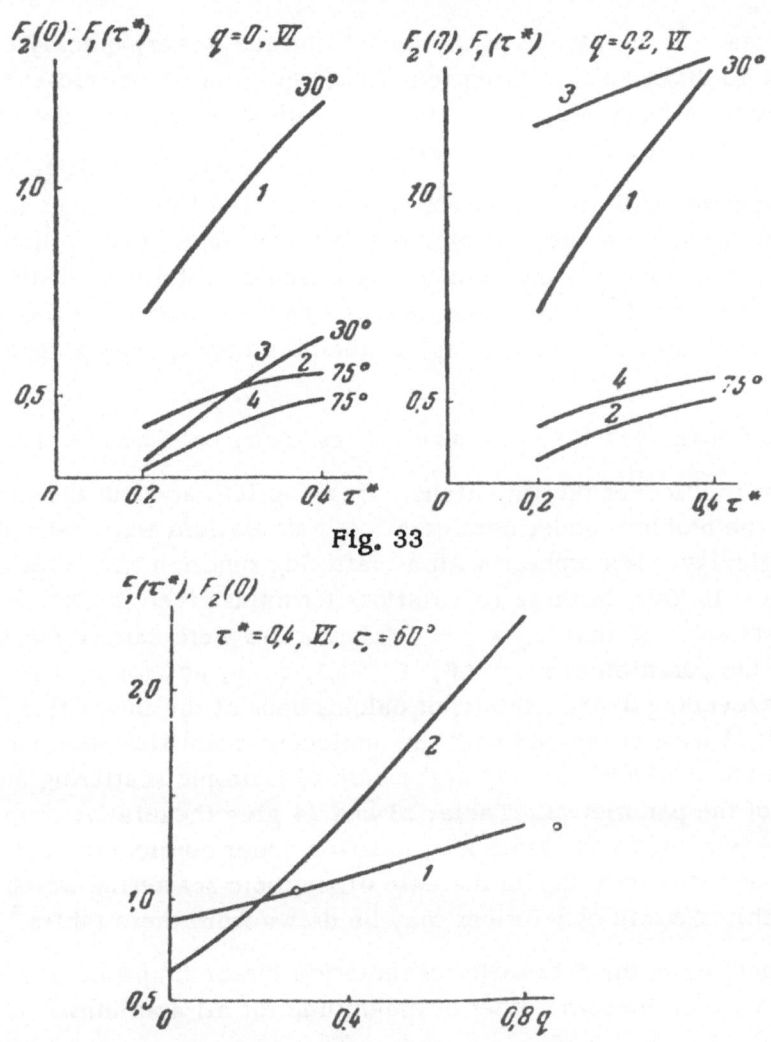

Fig. 33

Fig. 34

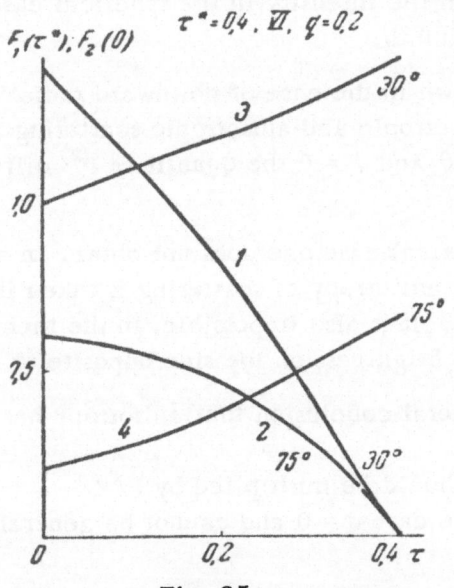

Fig. 35

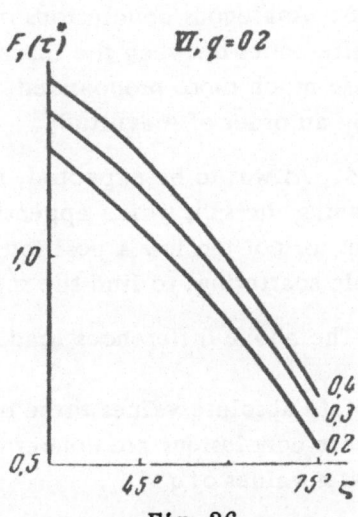

Fig. 36

An increase in the albedo of the earth's surface $\underline{q}$ very quickly equalizes fluxes corresponding to different scattering functions, and brings them closer to the case of spherical scattering functions, so that when the albedo is of the order of 0.3-0.4, the effect of anisotropy ceases to be important.

In comparison with the above results on the role of the albedo and the anisotropy of the scattering function, the new element in these conclusions is that the downward flux of scattered radiation is less sensitive to changes in the form of the scattering function than the intensity $I^{(2)}$. As $\underline{q}$ increases, the dependence of the flux on the degree of anisotropy decreases much more rapidly than the corresponding dependence of the intensity.

§ 5. Comparison with the Case of Isotropic Scattering

In order to discover the importance of taking into account the anisotropy of scattering in the problem under consideration, calculations were carried out in the case of a single-layer atmosphere with a scattering function characterized by $\Gamma_1/\Gamma_2 = = 10.7$ and $S_h = 10$ km. In these calculations formulas given in Chapter I were used, in which it was assumed that $\tau_1 = \tau*$. Calculations were carried out for the following values of the parameters: $\tau* = 0.6$, $\zeta = 60°$, $q = 0$, and for $\psi = 90$ and $180°$, with an error not exceeding 2-3%. Results of calculations of the intensities $I^{(1)}(\tau, \theta, \psi)$ and $I^{(2)}(\tau, \theta, \psi)$ were compared with the analogous quantities obtained by E. S. Kuznetsov and B. V. Ovchinskii [1] in the case of isotropic scattering and with the same values of the parameters. Tables 13 and 14 give the relative magnitudes of $I^{(1)}$ and $I^{(2)1}$ corresponding to the scattering function under consideration (the quantities in brackets represent the intensity in the case of isotropic scattering which is independent of the azimuth). Certain conclusions may be drawn from these tables.[2]

1. In both cases the intensities of radiation directed upward at small angles ($0 \leq \theta \leq 45°$) are of the same order of magnitude for all azimuths.

2. The intensity of radiation directed upward at larger angles θ in the case of nonspherical scattering is very different from the intensity in the spherical case and this difference increases with decreasing azimuth.

3. Analogous conclusions may be drawn in the case of downward radiation. The differences between the intensities in isotropic and anisotropic scattering are in this case much more pronounced. At large θ and $\psi = 0$ the quantities $I^{(2)}$ differ almost by an order of magnitude.

4. As was to be expected, in the spherical case one does not obtain an aureole surrounding the sun, which appears when the anisotropy of scattering is taken into account for not too low a position of the sun. It is also impossible, in the theory of isotropic scattering, to find the minimum of brightness on the side opposite to the sun.

The above differences lead to the general conclusion that anisotropy has its

---

[1] To obtain absolute values these quantities should be multiplied by $S/2$.
[2] All these conclusions are concerned with the case $q = 0$ and cannot be generalized to greater values of $q = 0$

TABLE 11

τ^*	q	No. of funct.	$\xi = 30°$		$\xi = 75°$	
			$F_1(\tau^*)$	$\delta, \%$	$F_1(\tau^*)$	$\delta, \%$
0,4	0	VI	0,6525	22,9	0,5055	9,4
		VII	0,5180		0,4599	
	0,2	VI	1,4182	5,7	0,6560	5,6
		VII	1,3389		0,6201	
	0,8	VI	4,5517	2,0	1,1950	1,1
		VII	4,5017		1,1818	
0,8	0	VII	0,9346	18,3	0,6283	7,5
		VIII	0,7776		0,5829	
	0,2	VII	1,5555	6,1	0,7218	4,5
		VIII	1,4639		0,6900	
	0,8	VII	3,9494	2,0	1,1279	5,7
		VIII	4,0296		1,0656	

most pronounced effect in the case of calculations of intensity of downward radiation, and also in the case of upward radiation for large values of θ.

§ 6. The Role of Multiple Scattering

The complexity of the problem of radiation transfer theory is associated with the necessity of taking into account multiple scattering of light. Taking into account only single scattering does not, as is clear from Eqs. (1.20)-(1.23), lead to any difficulties for any form of the scattering function. It is natural to try and explain what is the role of multiple scattering of radiation under different conditions of propagation of light in the atmosphere, and under what conditions can it be neglected. For this purpose a comparison has been carried out of the intensity of upward radiation for $\tau = \tau^*$, and of downward radiation for $\tau = 0$, with the corresponding intensities obtained by taking into account only single scattering from the formulas

$$I^{(j)}(\tau, 0, \psi) = \left[\frac{a_0^{(j)}}{2} + \sum_{k=1}^{10} a_k^{(j)}(\tau, 0) \cos k\psi \right] e^{-\tau^* \sec \zeta} (j = 1, 2).$$

The quantities $a_k^{(j)}(\tau, \theta)$ are determined using (1.20)-(1.23). Table 15 gives the relative errors (in %) which are admissible in the determination of $I^{(1)}(\tau^*, \theta, \psi)$

TABLE 12

τ°	q	No. of funct.	$\xi=30°$ $F_2(\theta)$	δ, %	$\xi=75°$ $F_2(\theta)$	δ, %
0,4	0	spherical	0,9814		0,5636	
				20,6		0
		VI	1,2073		0,5637	
				13,6		5,8
		VII	1,3838		0,5973	
	0,2	spherical	1,2177		0,6095	
				14,3		1,2
		VI	1,4048		0,6025	
				3,7		1,2
		VII	1,4487		0,6099	
	0,8	spherical	2,1086		0,7823	
				4,7		5,4
		VI	2,2123		0,7414	
				1,0		1,4
		VII	2,2341		0,7544	
0,8	0	VII	2,0761		0,5657	
				10,2		3,8
		VIII	2,2985		0,5876	
	0,2	VII	2,3241		0,6031	
				8,4		3,1
		VIII	2,5280		0,6219	
	0,8	VII	3,2820		0,7655	
		VIII		3,2		2,1
			3,3881		0,7494	

TABLE 13

ψ,°	0			90			180		
θ,° $\diagdown$ τ	0	45	75	0	45	75	0	45	75
0	0	0	0	0	0	0	0	0	0
0,2	0,0705 (0,0840)	0,1430 (0,1060)	0,6740 (0,2340)	0,0705	0,1083	0,2832	0,0705	0,1132	0,3112
0,4	0,1689 (0,1956)	0,3187 (0,3234)	1,1800 (0,4244)	0,1689	0,2510	0,5459	0,1689	0,2609	0,5826
0,6	0,2988 (0,3324)	0,5220 (0,3464)	1,5875 (0,5922)	0,2983	0,4300	0,8070	0,2988	0,4440	0,8405

TABLE 14

$\psi,°$ θ,° τ	0			90			180		
	0	45	75	0	45	75	0	45	75
0	0,4492 (0,3304)	2,1631 (0,3094)	3,4122 (0,4428)	0,4492	0,3562	0,5331	0,4492	0,2745	0,4716
0,2	0,3498 (0,2506)	1,8706 (0,2732)	3,5520 (0,4792)	0,3498	0,2908	0,5212	0,3498	0,2094	0,4671
0,4	0,2031 (0,1406)	1,2120 (0,1738)	2,7904 (0,3776)	0,2031	0,1644	0,3787	0,2031	0,1152	0,3322
0,6	0	0	0	0	0	0	0	0	0

and $I^{(2)}$ $(0, \theta, \psi)$ in taking into account only single scattering. These errors are de-
noted by p_1 and p_2.

As was to be expected the errors increase with increasing τ^* and reach 70%
at $\tau^* = 0.8$. However, even for a not very turbid atmosphere the error may exceed
25%, which shows that it is necessary to take into account multiple scattering in these
cases. The fraction due to multiple scattering in the intensities of downward and upward radiation is roughly the same, except in the cases where the directions of rays
are close to the direction toward the sun. In these cases the fraction of multiply
scattered radiation in the upward component remains roughly the same as in other
directions, but in the downward component the role of multiple scattering becomes
very small.[1] The latter fact appears to us to be quite natural and connected with the
nonspherical form of the scattering function. In directions close to the direction toward the sun, the quantity of single scattered radiation is so great that the role of
higher-order scattering is reduced.

The fraction of multiply scattered upward radiation increases with the azimuth
of the rays and reaches maximum values on the side opposite to the sun.

It should be noted that in the case of very turbid atmosphere the fraction of
multiply scattered radiation propagated in directions close to the direction toward
the sun is still sufficiently large, and in this case, for large zenith distances of the
sun, the errors p_1 and p_2 are close to each other.

As the zenith distance increases, no regular changes in the fraction of multiply
scattered radiation are found, and this apparently is connected with the mutually ex-
cluding influence of increasing amount of scattered light and attenuation due to an
increase in the path length of the ray.

No noticeable changes with the degree of isotropy of the scattering function

---

[1] For large zenith distances of the sun in these directions a reduction in the relative
error is also obtained for the intensity of upward radiation.

TABLE 15

τ^*	No. of funct	Error	θ°	$\psi=0°$; $\zeta=30°$						$\psi=0°$; $\zeta=75°$					
				0	15	30	45	60	75	0	15	30	45	60	75
0,2	V	p_1		16	19	20	21	22	20	21	21	18	17	13	11
	VI			15	20	17	21	24	19	20	22	18	17	13	10
	V	p_2		11	6	5	6	11	15	20	18	14	11	7	5
	VI			10	5	4	6	10	16	20	18	14	10	6	1
0,8	VII	p_1		40	48	48	50	52	44	45	44	43	39	33	24
	VIII			36	46	43	46	51	42	41	43	40	39	33	25
	VII	p_2		33	21	18	24	40	53	58	54	46	40	29	27
	VIII			35	21	17	24	40	58	60	56	48	41	16	28

$\tau°$	No. of funct	Error	θ°	$\psi=180°$; $\xi=30°$:						$\psi=180°$; $\xi=75°$;					
				0	15	30	45	60	75	0	15	30	45	60	75
0,2	V	p_1		16	12	13	15	17	21	21	20	19	18	16	15
	VI			15	14	15	16	17	22	20	20	21	18	17	18
	V	p_2		11	14	19	22	27	25	20	23	24	24	24	22
	VI			10	15	20	22	32	29	20	26	25	27	20	20
0,8	VII	p_1		40	38	39	41	41	44	45	44	42	36	35	31
	VIII			36	32	34	37	36	40	41	40	38	34	33	32
	VII	p_2		33	44	54	61	71	70	58	63	63	64	67	71
	VIII			35	47	59	68	75	75	60	65	64	66	70	76

are found either. Only at $\theta = \zeta$ the error p_2 decreases somewhat in accordance with the above remarks. Thus multiple scattering plays an important role in problems of transfer of short-wavelength radiation in all cases except for directions of propagation coinciding with the direction toward the sun.

§ 7. Explanation of Tables

Some results of calculations are published in the present paper. All the data are expressed in relative units. If the solar constant is put equal to πS, then absolute values may be obtained by multiplying the data given in the Tables by S/2. If σ (z) and the solar constant are taken for a given wavelength, then the data given in the Tables give the intensity of monochromatic radiation.

By putting π S equal to the integral solar constant, and taking σ (z) equal to the scattering coefficient averaged over the spectrum, the results of these calculations may be looked upon as the integral intensities of scattered radiation in the atmosphere with σ (z) independent of wavelength.

Since both in the first and in the second cases the model of the atmosphere considered here is characterized by the same scattering function independent of height, in each of the layers $0 \leq \tau \leq \tau_1, \tau_1 \leq \tau \leq \tau^*$.

The numerical material is divided into five Tables. Table I gives the intensities of scattered radiation $I^{(1)}$ (τ, θ, ψ) and $I^{(2)}$ (τ, θ, ψ) for q = 0, ζ = 30, 45, 60, and 75° and the following combinations of τ^* and the scattering function:

τ^*	0,2	0,4	0,6	0,8
No. of funct.	V, VI	VI, VII	VII, VIII	VII, VIII.

Table II gives solutions of the auxiliary equations which give $B_0^{(1)}$ (τ, θ) and $B_0^{(2)}$(τ, θ) [cf. (1.43)-(1.51)] for the same values of the parameters.

Table III contains the constants $\dfrac{1}{2} C \cdot e^{-\tau^* \sec \zeta} = C^*$ [cf. (1.52)]. Calculations of the intensity of scattered radiation for nonzero values of the albedo of the earth's surface $I_q^{(i)}(\tau$, θ, ψ), are carried out using

$$I_q^{(i)} (\tau, \theta, \psi) = I^{(i)} (\tau, \theta, \psi) + C^* B_0^{(i)} (\tau, \theta) \quad (i = 1, 2). \tag{3.7}$$

Table IV gives the values of the illumination of the earth's surface, i.e., the total downward flux of direct and scattered radiation at z = 0. This flux for q = 0 is calculated from

$$\bar{F}^{(2)} (0) = \int_0^{2\pi} \int_0^{\pi/2} I^{(2)} (0, \theta, \psi) \cos \theta \sin \theta \, d\theta \, d\psi + 2\pi \cos \zeta \, e^{-\tau^* \sec \zeta}. \tag{3.8}$$

The quantity

$$F^{(2)} (0) = \int_0^{2\pi} \int_0^{\pi/2} I^{(2)} (0, \theta, \psi) \cos \theta \sin \theta \, d\theta,$$

which represents the downward flux of scattered radiation at z = 0 is given separately, as well as the quantity $\pi \displaystyle\int_0^{\pi/2} B_0^{(2)}(0, \theta,) \sin 2\theta \, d\theta$ which is necessary in the calculation of the flux $\bar{F}_q^{(2)}(0)$ using the expression

$$F_q^{(2)}(0) = \bar{F}^{(2)}(0) + \pi C^* \int_0^{\pi/2} B_0^{(2)}(0, \theta) \sin 2\theta \, d\theta.$$

The transition from optical thicknesses τ to heights $\underline{z}$ may be carried out using Table V.

Practical use of Tables I-V requires a knowledge of the optical parameters τ^*, $\sigma(0)$ or S_h, q, ζ and the scattering function of the lower layers of the atmosphere. At the present time, however, sufficiently reliable and accurate determinations of all these quantities are not available.

At the same time, it is always possible to give rough values of τ^* and S_h, the average values of the albedo of the earth's surface $\underline{q}$ are known, and ζ may easily be determined.

The existence of these data, even when the scattering function is unknown, may, in our opinion, be used to obtain from Tables I and V useful information on the scattering of solar radiation in the atmosphere. The uncertainty introduced by the fact that the scattering function is unknown, and that the parameters τ^*, $\sigma(0)$, $\underline{q}$ are only roughly given, is partly removed with the help of the regularities in the behavior of the intensity as a function of each of the parameters which were found above.

CONTENTS OF TABLES

$\tau^*=0,2$; funct. V; $\psi=0°$ $\qquad\qquad$ $I^{(1)}(\tau, \theta, \psi)$

ζ,°	30						45					
θ, τ	0	15	30	45	60	75	0	15	30	45	60	75
0	0	0	0	0	0	0	0	0	0	0	0	0
0,05	0,0205	0,0175	0,0195	0,0229	0,0313	0,0721	0,0173	0,0151	0,0178	0 0221	0,0348	0,0831
0,1	0,0411	0,0349	0,0388	0,0454	0,0608	0,1344	0,0348	0,0302	0,0358	0,0433	0,0684	0,1561
0,125	0,0514	0,0438	0,0484	0,0562	0,0751	0,1627	0,0437	0.0380	0,0447	0,0541	0,08-6	0.1895
0,15	0,0617	0,0524	0,0581	0,0672	0,0887	0,1886	0,0525	0,0455	0,0536	0,0647	0,1C05	0,2208
0,16	0,0670	0,0568	0,0624	0,0722	0,0953	0,1991	0,0568	0,0496	0,0575	0,0698	0,1073	9,2333
0,17	0,0725	0,0614	0,0670	0,0772	0,1023	0,2089	0,0611	0,0536	0,0617	0.0747	0,1142	0,2456
0,18	0,0778	0,0657	0,0714	0,0818	0,1087	0,2185	0,0654	0,0576	0,0659	0,0796	0,1210	0,2578
0,19	0,0832	0,0701	0,0758	0,0866	0,1150	0,2281	0,0696	0,0616	0,0700	0,0847	0,1280	0,2693
0,195	0,0860	0,0725	0,0780	0,0892	0,1182	0,2327	0,0718	0,0637	0,0720	0,0871	0,1314	0,2753
0,197	0,0871	0,0732	0,0790	0,0899	0,1195	0,2346	0,0727	0,0645	0,0728	0,0882	0,1327	0,2776
0,2	0,0886	0,0748	0,0803	0,0914	0,1216	0,2373	0,0740	0,0657	0,0743	0,0897	0,1347	0,2811

$I^{(2)}(\tau, \theta, \psi)$

τ	0	15	30	45	60	75	0	15	30	45	60	75
0	0,1724	0,3149	0,4596	0,4111	0,3206	0,3734	0,1231	0,1772	0,3336	0,5373	0,5426	0,5012
0,05	0,1329	0,2407	0,3474	0,3151	0,2529	0,3122	0,0958	0,1377	0,2559	0,4115	0,4247	0,4164
0,1	0,0892	0,1557	0,2211	0,2072	0,1743	0,2283	0,0657	0,0932	0,1681	0,2655	0,2839	0,3022
0,125	0,0666	0,1100	0,1530	0,1475	0,1307	0,1770	0,0498	0,0691	0,1200	0,1850	0,2038	0,2323
0,15	0 0423	0 0625	0,0808	0,0846	0,0840	0,1215	0,0329	0,0439	0,C689	0,0984	0,1178	0,1547
0,16	0,0347	0,0516	0,0654	0,0677	0,C680	0,0993	0,0266	0,0355	0,0552	0,0798	0,0958	0,1266
0,17	0,0254	0,0374	0,0495	0,0519	0,0518	0,0760	0,0200	0,0267	0,0422	0,0613	0,0727	0,0973
0,18	0,0173	0,0252	0,0334	0,0350	0,0351	0,0520	0,0135	0,0181	0,0287	0,0414	0,0492	0,0664
0,19	0,0080	0,0120	0,0167	0,0178	0,0179	0,0265	0,0067	0,0089	0,0145	0,0210	0,0252	0,0341
0,195	0,0046	0,0067	0,0086	0,0090	0,0089	0,0131	0,0036	0,0047	0,0074	0,0107	0,0124	0,0171
0,197	0,0028	0,0044	0,0051	0,0048	0,0052	0,0080	0,0020	0,0026	0,0039	0,0062	0,0077	0,0003
0,2	0	0	0	0	0	0	0	0	0	0	0	0

$\tau^*=0,2$; funct. V; $\psi=45°$ $\qquad\qquad$ $I^{(1)}(\tau, \theta, \psi)$

τ	0	15	30	45	60	75	0	15	30	45	60	75
0	0	0	0	0	0	0	0	0	0	0	0	0
0,05	0,0205	0,0184	0,0197	0,0232	0,0303	0,0662	0,0173	0,0155	0,0181	0,0212	0,0317	0,0705
0,1	0,0411	0,0368	0,0391	0,0460	0,0594	0,1237	0,0348	0,0311	0,0363	0,0420	0,0621	0,1325
0,125	0,0514	0,0461	0,0487	0,0572	0,0732	0,1492	0,0437	0,0391	0,0455	0,0526	0,0768	0,1605
0,15	0,0617	0,0552	0,0585	0,0683	0,0867	0,1730	0,0525	0,0471	0,0545	0,0629	0,0912	0,1869
0,16	0,0670	0,0599	0,0629	0,0733	0,0933	0,1833	0,0568	0,0512	0,0586	0,0677	0,0978	0,1975
0,17	0,0725	0,0645	0,0676	0,0785	0,0999	0,1930	0,0611	0,0553	0,0629	0,0728	0,1045	0,2079
0,18	0,0778	0,0690	0,0722	0,0834	0.1062	0,2025	0,0654	0,0594	0,0672	0,0777	0,1109	0,2178
0,19	0,0832	0,0737	0,0767	0,0885	0,1126	0,2119	0,0696	0,0635	0,0713	0,0824	0,1176	0,2276
0,195	0,0860	0,0761	0,0788	0,0909	0,1157	0,2165	0,0718	0,0656	0,0735	0,0847	0,1210	0,2327
0,197	0,0871	0,0769	0,0798	0,0918	0,1170	0,2184	0,0727	0,0664	0,0743	0,0857	0,1223	0,2346
0,200	0,0886	0,0784	0,0813	0,0933	0,1189	0,2210	0,0740	0,0676	0,0756	0,0872	0,1240	0,2374

$I^{(2)}(\tau, \theta, \psi)$

τ	0	15	30	45	60	75	0	15	30	45	60	75
0	0,1724	0,2412	0,2602	0,2326	0,2401	0,3009	0,1231	0,1533	0,1889	0,2211	0,2589	0,3466
0,05	0,1329	0,1858	0,2000	0,1812	0,1952	0,2536	0,0958	0,1194	0,1471	0,1736	0,2063	0,2893
0,1	0,0892	0,1225	0,1328	0,1237	0,1357	0,1881	0,0657	0,0813	0,1005	0,1190	0,1438	0,2131
0,125	0,0666	0,0886	0,0968	0,0924	0,1028	0,1481	0,0498	0,0607	0,0751	0,0893	0,1089	0,1666
0,15	0,0423	0,0536	0,0590	0,0595	0,0679	0,1053	0,0329	0,0392	0,0485	0,0579	0,0718	0,1160
0,16	0,0356	0,0441	0,0477	0,0476	0,0549	0,0860	0,0266	0,0316	0,0388	0,0468	0,0582	0,0950
0,17	0,0254	0,0319	0,0360	0,0364	0,0418	0,0659	0,0200	0,0239	0,0296	0,0360	0,0443	0,0728
0,18	0,0173	0,0216	0,0243	0,0246	0,0282	0,0449	0,0135	0,0160	0,0201	0,0244	0,0299	0,0498
0,19	0,0080	0,0103	0,0121	0.0125	0,0144	0,0230	0,0067	0,0081	0,0102	0,0124	0,0153	0,0254
0,195	0,0046	0,0057	0,0062	0,0062	0,0072	0,0115	0,0036	0,0042	0,0052	0,0062	0,0076	0,0128
0,197	0,0028	0,0037	0,0037	0,0034	0,0042	0,0069	0,0020	0,0024	0,0028	0,0036	0,0047	0,0078
0,200	0	0	0	0	0	0	0	0	0	0	0	0

Table I 59

TABLE I

60						75						ζ,° θ,° τ
0	15	30	45	60	75	0	15	30	45	60	75	
0	0	0	0	0	0	0	0	0	0	0	0	0
0,0144	0,0144	0,0157	0,0226	0,0397	0,1029	0,0103	0,0109	0,0144	0,0213	0,0408	0,1048	0,05
0,0292	0,0291	0,0319	0,0454	0,0789	0,1970	0,0219	0,0231	0,0306	0,0450	0,0851	0,2110	0,1
0,0370	0,0368	0,0403	0,0571	0,0986	0,2413	0,0283	0,0297	0,0395	0,0578	0,1090	0,2654	0,125
0,0448	0,0446	0,0485	0,0687	0,1179	0,2837	0,0350	0,0367	0,0488	0,0713	0,1341	0,3216	0,15
0,0486	0,0482	0,0527	0,0738	0,1264	0,2982	0,0383	0,0402	0,0530	0,0772	0,1433	0,3387	0,16
0,0524	0,0518	0,0568	0,0790	0,1346	0,3124	0,0417	0,0437	0,0571	0,0833	0,1529	0,3562	0,17
0,0564	0,0556	0,0610	0,0844	0,1430	0,3262	0,0451	0,0473	0,0613	0,0896	0,1627	0,3741	0,18
0,0602	0,0594	0,0653	0,0896	0,1510	0,3401	0,0486	0,0511	0,0657	0,0958	0,1725	0,3925	0,19
0,0622	0,0612	0,0674	0,0923	0,1553	0,3472	0,0505	0,0530	0,0679	0,0990	0,1777	0,4016	0,195
0,0629	0,0600	0,0681	0,0932	0,1569	0,3497	0,0513	0,0538	0,0688	0,1004	0,1798	0,4054	0,197
0,0641	0,0631	0,0694	0,0948	0,1593	0,3539	0,0522	0,0548	0,0700	0,1023	0,1828	0,4111	0,2

60						75						τ
0,0857	0,1199	0,1823	0,3786	0,6771	0,8296	0,0596	0,0746	0,1096	0,1785	0,4155	0,9040	0
0,0681	0,0950	0,1439	0,2965	0,5341	0,6889	0,0491	0,0617	0,0905	0,1480	0,3454	0,7816	0,05
0,0483	0,0660	0,0993	0,1984	0,3551	0,4885	0,0362	0,0455	0,0662	0,1079	0,2458	0,5697	0,1
0,0375	0,0509	0,0748	0,1428	0,2507	0,3609	0,0286	0,0361	0,0518	0,0835	0,1831	0,4204	0,125
0,0262	0,0336	0,0481	0,0828	0,1357	0,2140	0,0203	0,0258	0,0355	0,0557	0,1100	0,2381	0,15
0,0212	0,0272	0,0389	0,0675	0,1106	0,1760	0,0166	0,0209	0,0290	0,0457	0,0903	0,1976	0,16
0,0160	0,0206	0,0298	0,0513	0,0844	0,1357	0,0127	0,0160	0,0223	0,0350	0,0696	0,1537	0,17
0,0108	0,0139	0,0201	0,0345	0,0573	0,0931	0,0086	0,0108	0,0151	0,0239	0,0478	0,1063	0,18
0,0055	0,0070	0,0103	0,0177	0,0292	0,0477	0,0044	0,0056	0,0077	0,0122	0,0245	0,0550	0,19
0,0028	0,0036	0,0051	0,0089	0,0148	0,0241	0,0022	0,0027	0,0039	0,0060	0,0122	0,0280	0,195
0,0017	0,0021	0,0029	0,0053	0,0087	0,0144	0,0013	0,0017	0,0024	0,0037	0,0074	0,0169	0,197
0	0	0	0	0	0	0	0	0	0	0	0	0,2

60						75						τ
0	0	0	0	0	0	0	0	0	0	0	0	0
0,0144	0,0146	0,0154	0,0204	0,0334	0,0715	0,0103	0,0106	0,0132	0,0181	0,0284	0,0641	0,05
0,0292	0,0297	0,0313	0,0412	0,0664	0,1365	0,0219	0,0225	0,0280	0,0381	0,0590	0,1282	0,1
0,0370	0,0374	0,0392	0,0517	0,0829	0,1667	0,0283	0,0289	0,0361	0,0489	0,0754	0,1606	0,125
0,0448	0,0453	0,0474	0,0623	0,0992	0,1956	0,0350	0,0358	0,0447	0,0603	0,0924	0,1936	0,15
0,0486	0,0489	0,0515	0,0672	0,1061	0,2079	0,0383	0,0393	0,0487	0,0654	0,0999	0,2069	0,16
0,0524	0,0527	0,0554	0,0723	0,1131	0,2199	0,0417	0,0427	0,0526	0,0704	0,1076	0,2203	0,17
0,0564	0,0564	0,0596	0,0773	0,1199	0,2317	0,0451	0,0462	0,0568	0,0755	0,1155	0,2339	0,18
0,0602	0,0602	0,0637	0,0823	0,1266	0,2435	0,0486	0,0501	0,0609	0,0809	0,1234	0,2477	0,19
0,0622	0,0622	0,0658	0,0849	0,1303	0,2492	0,0505	0,0519	0,0631	0,0837	0,1276	0,2547	0,195
0,0629	0,0629	0,0666	0,0859	0,1315	0,2517	0,0513	0,0527	0,0639	0,0847	0,1292	0,2574	0,197
0,0641	0,0641	0,0678	0,0870	0,1337	0,2551	0,0522	0,0537	0,0652	0,0863	0,1317	0,2616	0,200

60						75						τ
0,0857	0,1088	0,1403	0,1815	0,2392	0,3577	0,0596	0,0688	0,0891	0,1243	0,1824	0,2977	0
0,0681	0,0863	0,1110	0,1444	0,1936	0,3024	0,0491	0,0569	0,0740	0,1035	0,1538	0,2634	0,05
0,0483	0,0604	0,0772	0,1007	0,1372	0,2248	0,0362	0,0421	0,0548	0,0763	0,1144	0,2050	0,1
0,0375	0,0470	0,0587	0,0765	0,1052	0,1758	0,0286	0,0334	0,0434	0,0600	0,0900	0,1639	0,125
0,0262	0,0315	0,0389	0,0504	0,0700	0,1210	0,0203	0,0239	0,0308	0,0417	0,0619	0,1155	0,15
0,0212	0,0256	0,0314	0,0409	0,0570	0,0992	0,0166	0,0194	0,0252	0,0341	0,0507	0,0956	0,16
0,0160	0,0193	0,0239	0,0312	0,0435	0,0764	0,0127	0,0148	0,0192	0,0261	0,0390	0,0742	0,17
0,0108	0,0130	0,0161	0,0211	0,0294	0,0524	0,0086	0,0100	0,0131	0,0178	0,0268	0,0512	0,18
0,0055	0,0066	0,0082	0,0108	0,0150	0,0267	0,0044	0,0051	0,0067	0,0091	0,0138	0,0264	0,19
0,0028	0,0033	0,0041	0,0053	0,0076	0,0135	0,0022	0,0025	0,0033	0,0045	0,0068	0,0135	0,195
0,0017	0,0019	0,0025	0,0033	0,0045	0,0081	0,0013	0,0016	0,0020	0,0027	0,0042	0,0082	0,197
0	0	0	0	0	0	0	0	0	0	0	0	0,200

$\tau^*=0,2;$funct. V; $\psi=90°$ $I^{(1)}(\tau,\ \theta,\ \psi)$

ζ°	30						45					
θ,° τ	0	15	30	45	60	75	0	15	30	45	60	75
0	0	0	0	0	0	0	0	0	0	0	0	0
0,05	0,0205	0,0208	0,0218	0,0241	0,0329	0,0599	0,0173	0,0177	0,0188	0,0225	0,0315	0,0575
0,1	0,0411	0,0415	0,0431	0,0477	0,0643	0,1118	0,0348	0,0354	0,0376	0,0448	0,0619	0,1077
0,125	0,0514	0,0519	0,0537	0,0593	0,0793	0,1349	0,0437	0,0443	0,0472	0,0559	0,0765	0,1304
0,15	0,0617	0,0623	0,0644	0,0709	0,0939	0,1562	0,0525	0,0534	0,0566	0,0668	0,0907	0,1516
0,16	0,0670	0,0676	0,0693	0,0763	0,1008	0,1662	0,0568	0,0577	0,0613	0,0720	0,0974	0,1616
0,17	0,0725	0,0727	0,0746	0,0819	0,1075	0,1761	0,0611	0,0620	0,0659	0,0775	0,1039	0,1714
0,18	0,0778	0,0780	0,0797	0,0874	0,1143	0,1853	0,0654	0,0663	0,0705	0,0826	0,1103	0,1809
0,19	0,0832	0,0833	0,0849	0,0930	0,1210	0,1947	0,0696	0,0709	0,0752	0,0879	0,1169	0,1901
0,195	0,0860	0,0860	0,0874	0,0956	0,1242	0,1991	0,0718	0,0730	0,0775	0,0905	0,1201	0,1947
0,197	0,0871	0,0870	0,0885	0,0966	0,1255	0,2010	0,0727	0,0739	0,0784	0,0917	0,1213	0,1966
0,200	0,0886	0,0887	0,0899	0,0984	0,1275	0,2036	0,0740	0,0751	0,0799	0,0932	0,1233	0,1992

$I^{(2)}(\tau,\ \theta,\ \psi)$

τ	0	15	30	45	60	75	0	15	30	45	60	75
0	0,1724	0,1640	0,1542	0,1510	0,1607	0,2265	0,1231	0,1228	0,1225	0,1252	0,1479	0,2215
0,05	0,1329	0,1273	0,1195	0,1181	0,1280	0,1913	0,0958	0,0959	0,0958	0,0991	0,1184	0,1850
0,1	0,0892	0,0860	0,0813	0,0821	0,0907	0,1417	0,0657	0,0660	0,0667	0,0697	0,0841	0,1375
0,125	0,0666	0,0641	0,0611	0,0626	0,0707	0,1114	0,0498	0,0500	0,0508	0,0540	0,0656	0,1087
0,15	0,0423	0,0413	0,0399	0,0425	0,0495	0,0800	0,0329	0,0333	0,0347	0,0375	0,0459	0,0786
0,16	0,0347	0,0340	0,0323	0,0341	0,0400	0,0652	0,0266	0,0269	0,0278	0,0304	0,0373	0,0643
0,17	0,0254	0,0246	0,0244	0,0260	0,0304	0,0500	0,0200	0,0201	0,0211	0,0232	0,0283	0,0492
0,18	0,0173	0,0166	0,0165	0,0175	0,0205	0,0339	0,0135	0,0136	0,0144	0,0157	0,0190	0,0335
0,19	0,0080	0,0079	0,0082	0,0089	0,0105	0,0175	0,0067	0,0068	0,0072	0,0081	0,0098	0,0170
0,195	0,0046	0,0043	0,0041	0,0045	0,0053	0,0087	0,0036	0,0035	0,0037	0,0040	0,0049	0,0086
0,197	0,0028	0,0029	0,0026	0,0025	0,0030	0,0053	0,0020	0,0020	0,0019	0,0024	0,0031	0,0052
0,200	0	0	0	0	0	0	0	0	0	0	0	0

$\tau^*=0,2;$funct. V; $\psi=135°$ $I^{(1)}(\tau,\ \theta,\ \psi)$

τ	0	15	30	45	60	75	0	15	30	45	60	75
0	0	0	0	0	0	0	0	0	0	0	0	0
0,05	0,0205	0,0239	0,0254	0,0289	0,0382	0,0625	0,0173	0,0204	0,0227	0,0279	0,0384	0,0634
0,1	0,0411	0,0480	0,0508	0,0573	0,0747	0,1167	0,0348	0,0412	0,0455	0,0557	0,0755	0,1192
0,125	0,0514	0,0600	0,0635	0,0715	0,0921	0,1410	0,0437	0,0516	0,0569	0,0696	0,0935	0,1445
0,15	0,0617	0,0719	0,0761	0,0852	0,1093	0,1633	0,0525	0,0620	0,0684	0,0831	0,1112	0,1679
0,16	0,0670	0,0781	0,0826	0,0926	0,1173	0,1747	0,0568	0,0671	0,0745	0,0903	0,1195	0,1798
0,17	0,0725	0,0841	0,0893	0,0999	0,1250	0,1858	0,0611	0,0721	0,0805	0,0975	0,1279	0,1913
0,18	0,0778	0,0900	0,0958	0,1071	0,1328	0,1965	0,0654	0,0770	0,0866	0,1045	0,1361	0,2025
0,19	0,0832	0,0961	0,1026	0,1143	0,1406	0,2070	0,0696	0,0322	0,0927	0,1117	0,1444	0,2136
0,195	0,0860	0,0991	0,1059	0,1178	0,1443	0,2121	0,0718	0,0847	0,0958	0,1152	0,1485	0,2187
0,197	0,0871	0,1003	0,1072	0,1192	0,1459	0,2141	0,0727	0,0857	0,0969	0,1166	0,1503	0,2209
0,200	0,0896	0,1021	0,1092	0,1214	0,1482	0,2171	0,0740	0,0872	0,0989	0,1188	0,1525	0,2240

$I^{(2)}(\tau,\ \theta,\ \psi)$

τ	0	15	30	45	60	75	0	15	30	45	60	75
0	0,1724	0,1367	0,1199	0,1142	0,1328	0,2008	0,1231	0,0976	0,0923	0,0988	0,1192	0,2056
0,05	0,1329	0,1060	0,0931	0,0893	0,1056	0,1701	0,0958	0,0773	0,0721	0,0778	0,0958	0,1721
0,1	0,0892	0,0721	0,0643	0,0621	0,0750	0,1273	0,0657	0,0542	0,0504	0,0545	0,0691	0,1283
0,125	0,0666	0,0543	0,0490	0,0479	0,0588	0,1015	0,0498	0,0417	0,0389	0,0422	0,0547	0,1020
0,15	0,0423	0,0359	0,0333	0,0330	0,0417	0,0744	0,0329	0,0290	0,0269	0,0296	0,0397	0,0747
0,16	0,0347	0,0294	0,0268	0,0265	0,0336	0,0608	0,0266	0,0235	0,0215	0,0239	0,0322	0,0609
0,17	0,0254	0,0213	0,0203	0,0202	0,0256	0,0466	0,0200	0,0176	0,0162	0,0182	0,0245	0,0466
0,18	0,0173	0,0145	0,0136	0,0137	0,0172	0,0309	0,0135	0,0118	0,0110	0,0123	0,0163	0,0318
0,19	0,0080	0,0070	0,0069	0,0068	0,0087	0,0161	0,0067	0,0058	0,0054	0,0062	0,0084	0,0162
0,195	0,0046	0,0038	0,0035	0,0035	0,0044	0,0073	0,0036	0,0030	0,0029	0,0032	0,0041	0,0079
0,197	0,0028	0,0025	0,0021	0,0018	0,0026	0,0048	0,0020	0,0017	0,0015	0,0017	0,0025	0,0048
0,2	0	0	0	0	0	0	0	0	0	0	0	0

Table I   61

TABLE I (continued)

60						75						ζ°
0	15	30	45	60	75	0	15	30	45	60	75	θ,° / τ
0	0	0	0	0	0	0	0	0	0	0	0	0
0,0144	0,0149	0,0166	0,0204	0,0284	0,0532	8,0103	0,0107	0,0119	0,0147	0,0211	0,0406	0,05
0,0292	0,0302	0,0337	0,0410	0,0563	0,1008	0,0219	0,0227	0,0252	0,0308	0,0436	0,0803	0,1
0,0370	0,0382	0,0425	0,0516	0,0703	0,1230	0,0283	0,0293	0,0326	0,0396	0,0554	0,1001	0,125
0,0448	0,0462	0,0514	0,0620	0,0839	0,1438	0,0350	0,0363	0,0402	0,0487	0,0676	0,1198	0,15
0,0486	0,0501	0,0557	0,0669	0,0904	0,1539	0,0383	0,0396	0,0439	0,0532	0,0737	0,1297	0,16
0,0524	0,0541	0,0599	0,0719	0,0966	0,1638	0,0417	0,0431	0,0477	0,0578	0,0798	0,1396	0,17
0,0564	0,0580	0,0642	0,0769	0,1032	0,1735	0,0451	0,0466	0,0517	0,0625	0,0862	0,1495	0,18
0,0602	0,0620	0,0684	0,0817	0,1094	0,1832	0,0486	0,0503	0,0557	0,0673	0,0925	0,1594	0,19
0,0622	0,0640	0,0706	0,0843	0,1126	0,1878	0,0505	0,0521	0,0578	0,0698	0,0958	0,1646	0,195
0,0629	0,0648	0,0715	0,0854	0,1138	0,1897	0,0513	0,0529	0,0585	0,0707	0,0971	0,1664	0,197
0,0641	0,0661	0,0728	0,0868	0,1156	0,1926	0,0522	0,0540	0,0597	0,0722	0,0990	0,1695	0,200
0,0857	0,0869	0,0914	0,1038	0,1329	0,2050	0,0596	0,0614	0,0673	0,0798	0,1052	0,1661	0
0,0681	0,0692	0,0728	0,0830	0,1078	0,1738	0,0491	0,0506	0,0556	0,0664	0,0888	0,1473	0,05
0,0483	0,0490	0,0519	0,0591	0,0775	0,1312	0,0362	0,0372	0,0410	0,0492	0,0669	0,1165	0,1
0,0375	0,0388	0,0402	0,0461	0,0608	0,1048	0,0286	0,0295	0,0326	0,0392	0,0538	0,0953	0,125
0,0262	0,0267	0,0283	0,0322	0,0426	0,0767	0,0203	0,0211	0,0233	0,0281	0,0390	0,0718	0,15
0,0212	0,0216	0,0229	0,0261	0,0347	0,0628	0,0166	0,0170	0,0190	0,0229	0,0319	0,0593	0,16
0,0160	0,0163	0,0174	0,0198	0,0263	0,0483	0,0127	0,0130	0,0145	0,0176	0,0245	0,0459	0,17
0,0108	0,0110	0,0117	0,0135	0,0178	0,0330	0,0086	0,0088	0,0099	0,0119	0,0168	0,0316	0,18
0,0055	0,0056	0,0060	0,0068	0,0091	0,0169	0,0044	0,0045	0,0050	0,0061	0,0086	0,0163	0,19
0,0028	0,0029	0,0029	0,0034	0,0046	0,0085	0,0022	0,0022	0,0026	0,0030	0,0042	0,0083	0,195
0,0017	0,0017	0,0018	0,0021	0,0027	0,0051	0,0013	0,0014	0,0015	0,0019	0,0026	0,0051	0,197
0	0	0	0	0	0	0	0	0	0	0	0	0,200
0	0	0	0	0	0	0	0	0	0	0	0	0
0,0144	0,0157	0,0193	0,0248	0,0342	0,0620	0,0103	0,0111	0,0124	0,0161	0,0244	0,0468	0,05
0,0292	0,0317	0,0392	0,0500	0,0681	0,1180	0,0219	0,0234	0,0263	0,0338	0,0507	0,0929	0,1
0,0370	0,0403	0,0494	0,0629	0,0852	0,1441	0,0283	0,0302	0,0338	0,0435	0,0646	0,1161	0,125
0,0448	0,0488	0,0597	0,0758	0,1018	0,1688	0,0350	0,0374	0,0418	0,0536	0,0789	0,1393	0,15
0,0486	0,0530	0,0648	0,0820	0,1102	0,1809	0,0383	0,0409	0,0459	0,0588	0,0861	0,1511	0,16
0,0524	0,0571	0,0697	0,0883	0,1183	0,1929	0,0417	0,0446	0,0502	0,0641	0,0936	0,1629	0,17
0,0564	0,0613	0,0747	0,0946	0,1189	0,2044	0,0451	0,0483	0,0546	0,0695	0,1010	0,1748	0,18
0,0602	0,0656	0,0798	0,1008	0,1348	0,2161	0,0486	0,0522	0,0590	0,0750	0,1086	0,1869	0,19
0,0622	0,0677	0,0823	0,1041	0,1390	0,2217	0,0505	0,0541	0,0613	0,0779	0,1125	0,1929	0,195
0,0629	0,0685	0,0833	0,1053	0,1405	0,2240	0,0513	0,0549	0,0621	0,0790	0,1141	0,1951	0,197
0,0641	0,0699	0,0847	0,1072	0,1429	0,2274	0,0522	0,0560	0,0635	0,0807	0,1164	0,1989	0,200
0,0857	0,0761	0,0755	0,0836	0,1179	0,1981	0,0596	0,0564	0,0599	0,0741	0,1018	0,1702	0
0,0681	0,0603	0,0599	0,0672	0,0954	0,1690	0,0491	0,0465	0,0496	0,0617	0,0864	0,1518	0,05
0,0483	0,0426	0,0422	0,0485	0,0693	0,1288	0,0362	0,0346	0,0369	0,0459	0,0658	0,1213	0,1
0,0375	0,0338	0,0333	0,0384	0,0549	0,1041	0,0286	0,0276	0,0296	0,0367	0,0775	0,1003	0,125
0,0262	0,0230	0,0236	0,0277	0,0393	0,0776	0,0203	0,0200	0,0216	0,0267	0,0395	0,0769	0,15
0,0212	0,0187	0,0190	0,0226	0,0321	0,0636	0,0166	0,0162	0,0177	0,0218	0,0323	0,0637	0,16
0,0160	0,0141	0,0146	0,0172	0,0243	0,0488	0,0127	0,0123	0,0134	0,0167	0,0247	0,0492	0,17
0,0108	0,0094	0,0097	0,0116	0,0164	0,0333	0,0086	0,0083	0,0090	0,0113	0,0169	0,0340	0,18
0,0055	0,0048	0,0049	0,0059	0,0082	0,0170	0,0044	0,0042	0,0046	0,0057	0,0086	0,0175	0,19
0,0028	0,0024	0,0025	0,0030	0,0042	0,0085	0,0022	0,0021	0,0023	0,0028	0,0042	0,0089	1,125
0,0017	0,0015	0,0014	0,0016	0,0025	0,0052	0,0013	0,0013	0,0013	0,0016	0,0026	0,0054	0,197
0	0	0	0	0	0	0	0	0	0	0	0	0,2

$\tau^* = 0,2$; funct. V; $\psi = 180°$ $I^{(1)}(\tau, \theta, \psi)$

$\zeta°$	30						45					
$\theta,°$ \ τ	0	15	30	45	60	75	0	15	30	45	60	75
0	0	0	0	0	0	0	0	0	0	0	0	0
0,05	0,0205	0,0269	0,0336	0,0351	0,0427	0,0675	0,0172	0,0217	0,027(	0,0404	0,0496	0,0741
0,1	0,0411	0,0541	0,0671	0,0697	0,0833	0,1257	0,0348	0,0437	0,055-	0,0807	0,0976	0,1392
0,125	0,0514	0,0677	0,0938	0,0869	0,1028	0,1519	0,0437	0,0548	0,0693	0,1008	0,1209	0,1687
0,15	0,0617	0,0811	0,1005	0,1037	0,1219	0,1762	0,0525	0,0669	0,0832	0,1208	0,1439	0,1964
0,16	0,0670	0,0873	0,1064	0,1115	0,1317	0,1883	0,0568	0,0716	0,0898	0,1277	0,1541	0,2119
0,17	0,0725	0,0930	0,1123	0,1196	0,1413	0,2004	0,0611	0,0771	0,0964	0,1344	0,1644	0,2274
0,18	0,0778	0,0989	0,1181	0,1270	0,1508	0,2117	0,0654	0,0826	0,1030	0,1414	0,1741	0,2421
0,19	0,0832	0,1048	0,1239	0,1349	0,1603	0,2227	0,0696	0,0883	0,1096	0,1482	0,1845	0,2569
0,195	0,0860	0,1079	0,1267	0,1389	0,164<	0,2283	0,0718	0,0910	0,1128	0,1516	0,1893	0,2639
0,197	0,0871	0,1091	0,1279	0,1404	0,1667	0,2305	0,0727	0,0921	0,1141	0,1530	0,1914	0,2669
0,2	0,0886	0,1108	0,1297	0,1425	0,1697	0,2336	0,0740	0,0937	0,1162	0,1552	0,1942	0,2711

$I^{(2)}(\tau, \theta, \psi)$

	30						45					
0	0,1724	0,1281	0,1091	0,1085	0,1239	0,2051	0,1231	0,0902	0,0876	0,0922	0,1196	0,2140
0,05	0,1329	0,0997	0,0848	0,0844	0,0990	0,1735	0,0958	0,0712	0,0681	0,0728	0,0958	0,1798
0,1	0,0892	0,0680	0,0587	0,0589	0,0710	0,1293	0,0657	0,0500	0,0474	0,0515	0,0689	0,1350
0,125	0,0666	0,0515	0,0451	0,0453	0,0561	0,1020	0,0498	0,0387	0,0366	0,0403	0,0546	0,1081
0,15	0,0423	0,0344	0,0308	0,0311	0,0407	0,0744	0,0329	0,0275	0,0252	0,0289	0,0396	0,0799
0,16	0,0356	0,0281	0,0248	0,0249	0,0329	0,0607	0,0266	0,0221	0,0202	0,0232	0,0320	0,0646
0,17	0,0254	0,0204	0,0187	0,0189	0,0248	0,0466	0,0200	0,0166	0,0152	0,0179	0,0242	0,0500
0,18	0,0173	0,0138	0,0126	0,0128	0,0167	0,0317	0,0135	0,0111	0,0103	0,0120	0,0163	0,0341
0,19	0,0080	0,0066	0,0063	0,0064	0,0085	0,0161	0,0067	0,0056	0,0052	0,0060	0,0083	0,0175
0,195	0,0046	0,0036	0,0032	0,0033	0,0043	0,0081	0,0036	0,0029	0,0027	0,0031	0,0043	0,0086
0,197	0,0028	0,0023	0,0019	0,0017	0,0024	0,0048	0,0020	0,0017	0,0015	0,0017	0,0026	0,0053
0,200	0	0	0	0	0	0	0	0	0	0	0	0

$\tau^* = 0,2$; funct. VI; $\psi = 0°$ $I^{(1)}(\tau, \theta, \psi)$

	30						45					
0	0	0	0	0	0	0	0	0	0	0	0	0
0,05	0,0178	0,0132	0,0161	0,0205	0,0251	0,0707	0,0127	0,0128	0,0157	0,0170	0,0345	0,0801
0,1	0,0357	0,0263	0,0321	0,0405	0,0486	0,1310	0,0257	0,0256	0,0318	0,0340	0,0678	0,1502
0,125	0,0447	0,0331	0,0402	0,0502	0,0599	0,1584	0,0322	0,0322	0,0397	0,0422	0,0837	0,1820
0,15	0,0536	0,0394	0,0478	0,0601	0,0708	0,1836	0,0388	0,0386	0,0476	0,0506	0,0995	0,2117
0,16	0,0590	0,0440	0,0522	0,0650	0,0777	0,1939	0,0431	0,0427	0,0517	0,0556	0,1064	0,2242
0,17	0,0644	0,0486	0,0569	0,0698	0,0844	0,2036	0,0474	0,0467	0,0556	0,0608	0,1130	0,2364
0,18	0,0698	0,0529	0,0613	0,0746	0,0912	0,2132	0,0518	0,0507	0,0598	0,0657	0,1197	0,2485
0,19	0,0751	0,0573	0,0656	0,0794	0,0978	0,2225	0,0561	0,0548	0,0639	0,0708	0,1264	0,2603
0,195	0,0778	0,0597	0,0679	0,0819	0,1011	0,2271	0,0583	0,0569	0,0658	0,0732	0,1297	0,2663
0,197	0,0790	0,0604	0,0688	0,0827	0,1023	0,2289	0,0591	0,0577	0,0667	0,0743	0,1311	0,2685
0,2	0,0806	0,0620	0,0702	0,0843	0,1044	0,2316	0,0605	0,0588	0,0682	0,0758	0,1329	0,2719

$I^{(2)}(\tau, \theta, \psi)$

	30						45					
0	0,2011	0,3927	0,5900	0,5164	0,3719	0,3929	0,1272	0,2035	0,4183	0,6878	0,6785	0,5832
0,05	0,1525	0,2951	0,4387	0,3888	0,2884	0,3222	0,0981	0,1556	0,3154	0,5181	0,5223	0,4812
0,1	0,0988	0,1838	0,2689	0,2461	0,1926	0,2325	0,0665	0,1025	0,1994	0,3220	0,3350	0,3376
0,125	0,0714	0,1239	0,1772	0,1671	0,1396	0,1783	0,0500	0,0736	0,1357	0,2140	0,2306	0,2505
0,15	0,0420	0,0620	0,0804	0,0841	0,0832	0,1200	0,0326	0,0436	0,0685	0,0978	0,1169	0,1532
0,16	0,0343	0,0513	0,0651	0,0673	0,0674	0,0981	0,0263	0,0353	0,0549	0,0794	0,0950	0,1254
0,17	0,0252	0,0372	0,0492	0,0515	0,0513	0,0751	0,0198	0,0265	0,0420	0,0610	0,0724	0,0964
0,18	0,0171	0,0250	0,0333	0,0348	0,0348	0,0514	0,0134	0,0180	0,0285	0,0411	0,0489	0,0658
0,19	0,0080	0,0120	0,0166	0,0178	0,0178	0,0262	0,0066	0,0088	0,0144	0,0208	0,0250	0,0338
0,195	0,0045	0,0066	0,0086	0,0089	0,0087	0,0130	0,0035	0,0046	0,0073	0,0105	0,0123	0,0169
0,197	0,0027	0,0044	0,0051	0,0048	0,0052	0,0079	0,0019	0,0026	0,0039	0,0061	0,0076	0,0102
0,2	0	0	0	0	0	0	0	0	0	0	0	0

Table I 63

TABLE I(continued)

	60						75					ζ,°
0	15	30	45	60	75	0	15	30	45	60	75	θ,° / τ
0	0.	0	0	0	0	0	0	0	0	0	0	0
0,0144	0,0164	0,0215	0,0320	0,0516	0,0831	0,0103	0,0111	0,0132	0,0187	0,0326	0,0719	0,05
0,0292	0,0334	0,0437	0,0646	0,1033	0,1585	0,0219	0,0236	0,0282	0,0394	0,0679	0,1441	0,1
0,0370	0,0422	0,0551	0,0812	0,1291	0,1939	0,0282	0,0304	0,0364	0,0506	0,086ь	0,1809	0,125
0,0448	0,0511	0,0667	0,0980	0,1548	0,2278	0,0350	0,0376	0,0450	0,0623	0,1066	0,2183	0,15
0,0486	0,0554	0,0728	0,1057	0,1638	0,2435	0,0383	0,0413	0,0495	0,0691	0,1162	0,2309	0,16
0,0524	0,0598	0,0789	0,1133	0,1725	0,2590	0,0417	0,0450	0,0540	0,0759	0,1258	0,2437	0,17
0,0564	0,0642	0,0848	0,1208	0,1813	0,2745	0,0451	0,0489	0,0587	0,0829	0,1357	0,2566	0,18
0,0602	0,0686	0,0909	0,1286	0,1901	0,2895	0,0486	0,0528	0,0635	0,0901	0,1458	0,2697	0,19
0,0622	0,0709	0,0939	0,1325	0,1947	0,2970	0,0505	0,0548	0,0659	0,0939	0,1509	0,2764	0,195
0,0629	0,0717	0,0952	0,1341	0,1965	0,3000	0,0513	0,0556	0,0669	0,0952	0,1529	0,2789	0,197
0,0641	0,0732	0,0970	0,1364	0,1990	0,3046	0,0522	0,0568	0,0683	0,0974	0,1561	0,2831	0,2

0	15	30	45	60	75	0	15	30	45	60	75	ζ,°
0,0857	0,0734	0,0705	0,0835	0,1221	0,2175	0,0596	0,0547	0,0610	0,0773	0,1119	0,2086	0
0,0681	0,0584	0,0562	0,0670	0,0997	0,1858	0,0491	0,0453	0,0504	0,0645	0,0954	0,1883	0,05
0,0483	0,0411	0,0404	0,0485	0,0729	0,1424	0,0362	0,0337	0,0374	0,0484	0,0729	0,1535	0,1
0,0375	0,0326	0,0320	0,0384	0,0580	0,1153	0,0286	0,0270	0,0299	0,0390	0,0592	0,1290	0,125
0,0262	0,0224	0,0232	0,0277	0,0422	0,0863	0,0203	0,0197	0,0217	0,0286	0,0439	0,1011	0,15
0,0212	0,0181	0,0186	0,0224	0,0344	0,0709	0,0166	0,0159	0,0175	0,0232	0,0361	0,0837	0,16
0,0160	0,0135	0,0142	0,0171	0,0260	0,0545	0,0127	0,0122	0,0134	0,0179	0,0276	0,0650	0,17
0,0108	0,0092	0,0095	0,0115	0,0176	0,0373	0,0086	0,0083	0,0091	0,0120	0,0190	0,0448	0,18
0,0055	0,0046	0,0049	0,0059	0,0089	0,0191	0,0044	0,0042	0,0046	0,0062	0,0097	0,0231	0,19
0,0028	0,0024	0,0024	0,0030	0,0045	0,0095	0,0022	0,0021	0,0023	0,0031	0,0048	0,0118	0,195
0,0017	0,0015	0,0014	0,0017	0,0027	0,0057	0,0013	0,0013	0,0014	0,0019	0,0030	0,0071	0,197
0	0	0	0	0	0	0	0	0	0	0	0	0,200

0	0	0	0	0	0	0	0	0	0	0	0	0
0,0120	0,0128	0,0125	0,0224	0,0383	0,1069	0,0092	0,0085	0,0139	0,0203	0,0420	0,1253	0,05
0,0243	0,0262	0,0251	0,0451	0,0763	0,2040	0,0196	0,0181	0,0297	0,0431	0,0879	0,2527	0,1
0,0306	0,0330	0,0315	0,0568	0,0952	0,2496	0,0252	0,0234	0,0383	0,0554	0,1126	0,3183	0,125
0,0372	0,0400	0,0382	0,0684	0,1140	0,2936	0,0313	0,0287	0,0474	0,0684	0,1386	0,3858	0,15
0,0410	0,0436	0,0423	0,0735	0,1223	0,3074	0,0346	0,0323	0,0513	0,0742	0,1477	0,4005	0,16
0,0448	0,0471	0,0465	0,0785	0,1307	0,3210	0,0380	0,0359	0,0555	0,0803	0,1570	0,4155	0,17
0,0487	0,0509	0,0508	0,0838	0,1388	0,3344	0,0414	0,0395	0,0597	0,0866	0,1667	0,4311	0,18
0,0526	0,0542	0,0549	0,0888	0,1468	0,3477	0,0449	0,0433	0,0640	0,0928	0,1765	0,4472	0,19
0,0546	0,0565	0,0572	0,0916	0,1511	0,3543	0,0467	0,0453	0,0663	0,0950	0,1715	0,4552	0,195
0,0554	0,0572	0,0580	0,0926	0,1530	0,3571	0,0475	0,0460	0,0672	0,0974	0,1835	0,4585	0,197
0,0565	0,0583	0,0593	0,0942	0,1552	0,3611	0,0485	0,0472	0,0685	0,0993	0,1867	0,4635	0,2

0,0843	0,1234	0,2099	0,4716	0,8658	1,0455	0,0594	0,0732	0,1132	0,2064	0,4771	1,1119	0
0,0667	0,0971	0,1633	0,3632	0,6719	0,8533	0,0488	0,0604	0,0931	0,1688	0,3922	0,9481	0,05
0,0472	0,0668	0,1096	0,2343	0,4304	0,5819	0,0357	0,0447	0,0673	0,1193	0,2724	0,6698	0,1
0,0369	0,0505	0,0798	0,1611	0,2898	0,4104	0,0282	0,0356	0,0522	0,0894	0,1971	0,4751	0,125
0,0260	0,0334	0,0478	0,0823	0,1349	0,2130	0,0202	0,0256	0,0353	0,0555	0,1096	0,2375	0,15
0,0210	0,0270	0,0387	0,0670	0,1099	0,1752	0,0165	0,0207	0,0288	0,0454	0,0901	0,1971	0,16
0,0159	0,0205	0,0296	0,0511	0,0839	0,1350	0,0126	0,0159	0,0222	0,0348	0,0694	0,1533	0,17
0,0107	0,0138	0,0200	0,0344	0,0570	0,0926	0,0086	0,0108	0,0150	0,0238	0,0477	0,1062	0,18
0,0054	0,0070	0,0102	0,0176	0,0290	0,0474	0,0044	0,0055	0,0077	0,0121	0,0244	0,0549	0,19
0,0028	0,0035	0,0050	0,0088	0,0148	0,0240	0,0021	0,0028	0,0039	0,0061	0,0123	0,0279	0,195
0,0016	0,0021	0,0029	0,0054	0,0086	0,0143	0,0014	0,0017	0,0024	0,0037	0,0073	0,0168	0,197
0	0	0	0	0	0	0	0	0	0	0	0	0,2

$\tau^{*}=0,2$;funct.VI; $\psi=45°$ $I^{(1)}(\tau, \theta, \psi)$

$r,°$	30						45					
$\theta,°$ \ τ	0	15	30	45	60	75	0	15	30	45	60	75
0	0	0	0	0	0	0	0	0	0	0	0	0
0,05	0,0178	0,0145	0,0148	0,0214	0,0250	0,0583	0,0127	0,0122	0,0166	0,0173	0,0270	0,0703
0,1	0,0357	0,0287	0,0295	0,0424	0,0485	0,1077	0,0257	0,0246	0,0335	0,0343	0,0527	0,1319
0,125	0,0447	0,0359	0,0368	0,0527	0,0597	0,1300	0,0322	0,0308	0,0418	0,0428	0,0649	0,1596
0,15	0,0536	0,0430	0,0440	0,0630	0,0704	0,1505	0,0388	0,0369	0,0503	0,0512	0,0772	0,1855
0,16	0,0590	0,0477	0,0487	0,0680	0,0773	0,1612	0,0431	0,0412	0,0544	0,0561	0,0841	0,1959
0,17	0,0644	0,0524	0,0532	0,0728	0,0839	0,1715	0,0474	0,0452	0,0584	0,0611	0,0908	0,2059
0,18	0,0698	0,0571	0,0578	0,0779	0,0904	0,1815	0,0518	0,0494	0,0627	0,0659	0,0974	0,2158
0,19	0,0751	0,0617	0,0625	0,0829	0,0968	0,1913	0,0561	0,0537	0,0670	0,0709	0,1041	0,2254
0,195	0,0778	0,0640	0,0648	0,0853	0,1000	0,1962	0,0583	0,0557	0,0690	0,0734	0,1075	0,2303
0,197	0,0790	0,0648	0,0658	0,0863	0,1014	0,1981	0,0591	0,0565	0,0697	0,0744	0,1089	0,2321
0,2	0,0806	0,0667	0,0671	0,0877	0,1033	0,2008	0,0605	0,0577	0,0710	0,0759	0,1107	0,2349

$I^{(2)}(\tau, \theta, \psi)$

τ	0	15	30	45	60	75	0	15	30	45	60	75
0	0,2011	0,2930	0,3186	0,2708	0,2600	0,3009	0,1272	0,1692	0,2193	0,2550	0,2825	0,3536
0,05	0,1525	0,2218	0,2402	0,2073	0,2034	0,2488	0,0981	0,1300	0,1679	0,1968	0,2221	0,2927
0,1	0,0988	0,1409	0,1535	0,1371	0,1393	0,1837	0,0665	0,0866	0,1110	0,1308	0,1515	0,2135
0,125	0,0714	0,0977	0,1071	0,0988	0,1041	0,1451	0,0500	0,0633	0,0803	0,0952	0,1126	0,1660
0,15	0,0420	0,0532	0,0586	0,0590	0,0671	0,1038	0,0326	0,0389	0,0481	0,0574	0,0710	0,1147
0,16	0,0343	0,0439	0,0473	0,0472	0,0543	0,0848	0,0263	0,0314	0,0384	0,0465	0,0577	0,0938
0,17	0,0252	0,0318	0,0357	0,0361	0,0413	0,0650	0,0198	0,0237	0,0294	0,0357	0,0440	0,0721
0,18	0,0171	0,0214	0,0241	0,0244	0,0279	0,0443	0,0134	0,0159	0,0198	0,0240	0,0297	0,0492
0,19	0,0080	0,0102	0,0121	0,0124	0,0143	0,0226	0,0066	0,0080	0,0101	0,0121	0,0152	0,0252
0,195	0,0045	0,0056	0,0062	0,0061	0,0070	0,0113	0,0035	0,0041	0,0051	0,0061	0,0075	0,0127
0,197	0,0027	0,0037	0,0037	0,0034	0,0042	0,0068	0,0019	0,0023	0,0029	0,0035	0,0046	0,0077
0,2	0	0	0	0	0	0	0	0	0	0	0	0

$\tau^{*}=0,2$;functVI; $\psi=90°$ $I^{(1)}(\tau, \theta, \psi)$

τ	0	15	30	45	60	75	0	15	30	45	60	75
0	0	0	0	0	0	0	0	0	0	0	0	0
0,05	0,0178	0,0179	0,0168	0,0178	0,0296	0,0534	0,0127	0,0129	0,0138	0,0188	0,0295	0,0500
0,1	0,0357	0,0356	0,0334	0,0351	0,0577	0,0988	0,0257	0,0257	0,0278	0,0376	0,0578	0,0933
0,125	0,0447	0,0446	0,0418	0,0436	0,0710	0,1190	0,0322	0,0323	0,0347	0,0468	0,0714	0,1125
0,15	0,0536	0,0533	0,0499	0,0519	0,0842	0,1377	0,0388	0,0388	0,0415	0,0560	0,0848	0,1304
0,16	0,0590	0,0587	0,0551	0,0576	0,0911	0,1480	0,0431	0,0432	0,0463	0,0613	0,0914	0,1410
0,17	0,0644	0,0640	0,0603	0,0632	0,0979	0,1584	0,0474	0,0477	0,0509	0,0666	0,0979	0,1514
0,18	0,0698	0,0693	0,0654	0,0689	0,1045	0,1679	0,0518	0,0520	0,0557	0,0718	0,1045	0,1614
0,19	0,0751	0,0744	0,0706	0,0745	0,1113	0,1777	0,0561	0,0566	0,0606	0,0773	0,1109	0,1711
0,195	0,0778	0,0771	0,0732	0,0773	0,1146	0,1822	0,0583	0,0587	0,0627	0,0798	0,1141	0,1761
0,197	0,0790	0,0781	0,0743	0,0784	0,1158	0,1842	0,0591	0,0597	0,0638	0,0811	0,1153	0,1779
0,200	0,0806	0,0798	0,0758	0,0800	0,1179	0,1867	0,0605	0,0610	0,0652	0,0826	0,1173	0,1807

$I^{(2)}(\tau, \theta, \psi)$

τ	0	15	30	45	60	75	0	15	30	45	60	75
0	0,2011	0,1870	0,1649	0,1504	0,1620	0,2279	0,1272	0,1253	0,1214	0,1244	0,1516	0,2161
0,05	0,1525	0,1428	0,1261	0,1166	0,1277	0,1874	0,0981	0,0970	0,0944	0,0977	0,1198	0,1788
0,1	0,0988	0,0935	0,0842	0,0806	0,0897	0,1377	0,0665	0,0663	0,0654	0,0682	0,0841	0,1323
0,125	0,0714	0,0676	0,0621	0,0616	0,0696	0,1081	0,0500	0,0499	0,0499	0,0529	0,0652	0,1053
0,15	0,0420	0,0410	0,0395	0,0419	0,0487	0,0784	0,0326	0,0330	0,0343	0,0370	0,0451	0,0774
0,16	0,0343	0,0337	0,0319	0,0337	0,0394	0,0640	0,0263	0,0266	0,0274	0,0300	0,0368	0,0632
0,17	0,0252	0,0245	0,0241	0,0257	0,0299	0,0490	0,0198	0,0199	0,0210	0,0230	0,0279	0,0485
0,18	0,0171	0,0164	0,0163	0,0173	0,0202	0,0333	0,0134	0,0135	0,0142	0,0154	0,0188	0,0330
0,19	0,0080	0,0078	0,0081	0,0089	0,0104	0,0171	0,0066	0,0067	0,0071	0,0079	0,0097	0,0168
0,195	0,0045	0,0043	0,0041	0,0044	0,0051	0,0086	0,0035	0,0034	0,0036	0,0039	0,0048	0,0085
0,197	0,0027	0,0028	0,0025	0,0023	0,0030	0,0052	0,0019	0,0020	0,0019	0,0024	0,0030	0,0051
0,2	0	0	0	0	0	0	0	0	0	0	0	0

Table I   65

TABLE I(continued)

60						75						ζ,°
0	15	30	45	60	75	0	15	30	45	60	75	θ,° / τ
0	0	0	0	0	0	0	0	0	0	0	0	0
0,0120	0,0134	0,0124	0,0174	0,0331	0,0707	0,0092	0,0083	0,0114	0,0178	0,0277	0,0627	0,05
0,0243	0,0273	0,0253	0,0348	0,0659	0,1343	0,0196	0,0176	0,0243	0,0376	0,0579	0,1253	0,1
0,0306	0,0345	0,0319	0,0437	0,0821	0,1640	0,0254	0,0225	0,0312	0,0483	0,0738	0,1569	0,125
0,0372	0,0418	0,0385	0,0527	0,0984	0,1921	0,0313	0,0278	0,0386	0,0596	0,0905	0,1889	0,15
0,0410	0,0455	0,0426	0,0577	0,1052	0,2042	0,0346	0,0313	0,0425	0,0646	0,0980	0,2023	0,16
0,0448	0,0491	0,0466	0,0627	0,1120	0,2162	0,0380	0,0349	0,0465	0,0696	0,1056	0,2157	0,17
0,0487	0,0529	0,0507	0,0678	0,1189	0,2279	0,0414	0,0385	0,0507	0,0747	0,1136	0,2294	0,18
0,0526	0,0567	0,0550	0,0727	0,1255	0,2395	0,0449	0,0422	0,0549	0,0799	0,1215	0,2432	0,19
0,0546	0,0585	0,0570	0,0754	0,1289	0,2452	0,0467	0,0441	0,0571	0,0827	0,1254	0,2502	0,195
0,0554	0,0593	0,0578	0,0764	0,1303	0,2478	0,0475	0,0448	0,0579	0,0838	0,1271	0,2529	0,197
0,0565	0,0604	0,0591	0,0780	0,1322	0,2511	0,0485	0,0461	0,0593	0,0854	0,1297	0,2571	0,2

60						75						ζ,°
0	15	30	45	60	75	0	15	30	45	60	75	θ,° / τ
0,0843	0,1084	0,1470	0,1971	0,2608	0,3757	0,0594	0,0686	0,0878	0,1259	0,2182	0,2816	0
0,0667	0,0855	0,1154	0,1550	0,2083	0,3144	0,0488	0,0564	0,0727	0,1043	0,1809	0,2489	0,05
0,0472	0,0598	0,0792	0,1061	0,1445	0,2301	0,0357	0,0417	0,0537	0,0764	0,1295	0,1952	0,1
0,0369	0,0459	0,0595	0,0789	0,1086	0,1779	0,0282	0,0332	0,0428	0,0599	0,0978	0,1581	0,125
0,0260	0,0312	0,0386	0,0500	0,0695	0,1200	0,0202	0,0237	0,0305	0,0415	0,0615	0,1149	0,15
0,0210	0,0253	0,0312	0,0406	0,0565	0,0984	0,0165	0,0192	0,0250	0,0339	0,0505	0,0952	0,16
0,0159	0,0191	0,0237	0,0310	0,0431	0,0758	0,0126	0,0147	0,0191	0,0260	0,0389	0,0738	0,17
0,0107	0,0129	0,0160	0,0210	0,0292	0,0519	0,0086	0,0100	0,0130	0,0178	0,0266	0,0509	0,18
0,0054	0,0065	0,0082	0,0108	0,0149	0,0265	0,0044	0,0050	0,0067	0,0090	0,0137	0,0263	0,19
0,0028	0,0033	0,0040	0,0052	0,0075	0,0134	0,0021	0,0026	0,0033	0,0046	0,0068	0,0134	0,195
0,0016	0,0019	0,0025	0,0033	0,0045	0,0080	0,0014	0,0016	0,0020	0,0027	0,0041	0,0081	0,197
0	0	0	0	0	0	0	0	0	0	0	0	0,2

60						75						ζ,°
0	15	30	45	60	75	0	15	30	45	60	75	θ,° / τ
0	0	0	0	0	0	0	0	0	0	0	0	0
0,0120	0,0126	0,0148	0,0190	0,0256	0,0446	0,0092	0,0095	0,0104	0,0124	0,0174	0,0335	0,05
0,0243	0,0257	0,0301	0,0381	0,0510	0,0841	0,0196	0,0203	0,0220	0,0262	0,0360	0,0659	0,1
0,0306	0,0323	0,0378	0,0478	0,0633	0,1022	0,0252	0,0261	0,0284	0,0335	0,0456	0,0817	0,125
0,0372	0,0392	0,0458	0,0576	0,0756	0,1192	0,0313	0,0323	0,0351	0,0412	0,0556	0,0973	0,15
0,0410	0,0431	0,0501	0,0624	0,0820	0,1300	0,0346	0,0357	0,0388	0,0457	0,0619	0,1079	0,16
0,0448	0,0470	0,0544	0,0673	0,0884	0,1405	0,0380	0,0391	0,0425	0,0505	0,0681	0,1185	0,17
0,0487	0,0511	0,0587	0,0721	0,0950	0,1509	0,0414	0,0427	0,0466	0,0552	0,0747	0,1290	0,18
0,0526	0,0551	0,0629	0,0770	0,1012	0,1610	0,0449	0,0463	0,05C6	0,0600	0,0811	0,1396	0,19
0,0546	0,0571	0,0651	0,0797	0,1044	0,1653	0,0467	0,0483	0,0527	0,0625	0,0844	0,1451	0,195
0,0554	0,0578	0,0659	0,0806	0,1058	0,1680	0,0475	0,0489	0,0535	0,0635	0,0857	0,1470	0,197
0,0565	0,0591	0,0673	0,0822	0,1075	0,1710	0,0485	0,0502	0,0547	0,0649	0,0877	0,1503	0,200

60						75						ζ,°
0	15	30	45	60	75	0	15	30	45	60	75	θ,° / τ
0,0843	0,0858	0,0916	0,1056	0,1349	0,1939	0,0594	0,0608	0,0661	0,0773	0,0994	0,1523	0
0,0667	0,0679	0,0724	0,0837	0,1081	0,1633	0,0488	0,0499	0,0546	0,0640	0,0838	0,1346	0,05
0,0472	0,0480	0,0512	0,0590	0,0769	0,1237	0,0357	0,0367	0,0402	0,0476	0,C636	0,1080	0,1
0,0369	0,0375	0,0398	0,0458	0,0600	0,100C	0,0282	0,0292	0,0320	0,0382	0,0516	0,0902	0,125
0,0260	0,0264	0,0280	0,0319	0,0421	0,075€	0,0202	0,0209	0,0230	0,0279	0,0386	0,0711	0,15
0,0210	0,0213	0,0227	0,0258	0,0342	0,062C	0,0165	0,0169	0,0188	0,0227	0,0316	0,0588	0,16
0,0159	0,0161	0,0172	0,0196	0,0260	0,0475	0,0126	0,0129	0,0144	0,0174	0,0243	0,0455	0,17
0,0107	0,0109	0,0116	0,0133	0,0176	0,0325	0,0086	0,0088	0,0099	0,0118	0,0166	0,0314	0,18
0,0054	0,0055	0,0059	0,0068	0,0090	0,0167	0,0044	0,0044	0,0050	0,0060	0,0085	0,0161	0,19
0,0028	0,0028	0,0029	0,0033	0,0045	0,C084	0,0021	0,0023	0,0025	0,0030	0,0043	0,C081	0,195
0,0016	0,0016	0,0017	0,0021	0,0027	0,0050	0,0014	0,0014	0,0015	0,0018	0,0025	0,0050	0,197
0	0	0	0	0	0	0	0	0	0	0	0	0,2

$\tau^*=0,2$; funct. VI; $\psi=135°$ $I^{(1)}(\tau,\theta,\psi)$

$\zeta,°$	30						45					
$\theta,°$ τ	0	15	30	45	60	75	0	15	30	45	60	75
0	0	0	0	0	0	0	0	0	0	0	0	0
0,05	0,0178	0,0197	0,0212	0,0254	0,0297	0,0494	0,0127	0,0170	0,0199	0,0244	0,0317	0,0486
0,1	0,0357	0,0392	0,0424	0,0502	0,0577	0,0922	0,0257	0,0341	0,0399	0,0487	0,0622	0,0906
0,125	0,0447	0,0491	0,0529	0,0625	0,0711	0,1109	0,0322	0,0428	0,0499	0,0607	0,0768	0,1092
0,15	0,0536	0,0589	0,0634	0,0745	0,0842	0,1283	0,0388	0,0513	0,0599	0,0727	0,0921	0,1263
0,16	0,0590	0,0650	0,0700	0,0819	0,0926	0,1405	0,0431	0,0564	0,0659	0,0790	0,0999	0,1397
0,17	0,0644	0,0710	0,0767	0,0891	0,1005	0,1526	0,0474	0,0615	0,0720	0,0870	0,1084	0,1523
0,18	0,0698	0,0769	0,0833	0,0964	0,1086	0,1642	0,0518	0,0663	0,0780	0,0941	0,1170	0,1646
0,19	0,0751	0,0830	0,0900	0,1037	0,1166	0,1756	0,0561	0,0717	0,0842	0,1013	0,1254	0,1767
0,195	0,0778	0,0860	0,0934	0,1072	0,1206	0,1810	0,0583	0,0742	0,0873	0,1047	0,1297	0,1828
0,197	0,0790	0,0872	0,0948	0,1087	0,1222	0,1833	0,0591	0,0753	0,0885	0,1062	0,1313	0,1852
0,200	0,0806	0,0891	0,0968	0,1108	0,1245	0,1865	0,0605	0,0767	0,0903	0,1084	0,1338	0,1885

$I^{(2)}(\tau,\theta,\psi)$

τ	0	15	30	45	60	75	0	15	30	45	60	75
0	0,2011	0,1467	0,1184	0,1168	0,1252	0,1780	0,1272	0,0979	0,0939	0,0941	0,1058	0,1990
0,05	0,1525	0,1124	0,0913	0,0900	0,0989	0,1485	0,0981	0,0764	0,0727	0,0736	0,0847	0,1650
0,1	0,0988	0,0750	0,0625	0,0618	0,0705	0,1132	0,0665	0,0532	0,0501	0,0515	0,0621	0,1229
0,125	0,0714	0,0554	0,0479	0,0473	0,0559	0,0930	0,0500	0,0411	0,0385	0,0405	0,0508	0,0982
0,15	0,0420	0,0354	0,0329	0,0326	0,0409	0,0730	0,0326	0,0288	0,0265	0,0292	0,0390	0,0734
0,16	0,0343	0,0291	0,0264	0,0261	0,0330	0,0596	0,0263	0,0231	0,0212	0,0235	0,0317	0,0600
0,17	0,0252	0,0211	0,0200	0,0199	0,0251	0,0456	0,0198	0,0174	0,0161	0,0181	0,0241	0,0460
0,18	0,0171	0,0142	0,0134	0,0135	0,0169	0,0312	0,0134	0,0118	0,0109	0,0121	0,0162	0,0312
0,19	0,0080	0,0069	0,0068	0,0069	0,0086	0,0158	0,0066	0,0059	0,0056	0,0061	0,0084	0,0160
0,195	0,0045	0,0037	0,0035	0,0034	0,0042	0,0080	0,0035	0,0030	0,0028	0,0031	0,0040	0,0080
0,197	0,0027	0,0024	0,0021	0,0020	0,0026	0,0047	0,0019	0,0017	0,0015	0,0018	0,0023	0,0048
0,2	0	0	0	0	0	0	0	0	0	0	0	0

$\tau^*=0,2$; funct. VI; $\psi=180°$ $I^{(1)}(\tau,\theta,\psi)$

τ	0	15	30	45	60	75	0	15	30	45	60	75
0	0	0	0	0	0	0	0	0	0	0	0	0
0,05	0,0178	0,0200	0,0229	0,0267	0,0373	0,0521	0,0127	0,0187	0,0210	0,0276	0,0378	0,0669
0,1	0,0357	0,0401	0,0456	0,0531	0,0724	0,0967	0,0257	0,0372	0,0420	0,0551	0,0743	0,1250
0,125	0,0447	0,0504	0,0570	0,0659	0,0894	0,1165	0,0322	0,0468	0,0526	0,0688	0,0920	0,1516
0,15	0,0536	0,0602	0,0683	0,0787	0,1062	0,1347	0,0388	0,0562	0,0632	0,0822	0,1093	0,1760
0,16	0,0590	0,0655	0,0744	0,0867	0,1161	0,1478	0,0431	0,0618	0,0698	0,0895	0,1198	0,1920
0,17	0,0644	0,0724	0,0805	0,0947	0,1256	0,1611	0,0479	0,0673	0,0764	0,0968	0,1307	0,2079
0,18	0,0698	0,0784	0,0866	0,1025	0,1352	0,1735	0,0518	0,0729	0,0830	0,1040	0,1411	0,2229
0,19	0,0751	0,0846	0,0924	0,1107	0,1447	0,1856	0,0561	0,0786	0,0898	0,1113	0,1518	0,2378
0,195	0,0778	0,0876	0,0957	0,1147	0,1494	0,1917	0,0583	0,0813	0,0931	0,1148	0,1568	0,2450
0,197	0,0790	0,0888	0,0969	0,1162	0,1513	0,1940	0,0591	0,0824	0,0944	0,1162	0,1590	0,2480
0,2	0,0806	0,0906	0,0989	0,1184	0,1542	0,1970	0,0605	0,0840	0,0964	0,1186	0,1619	0,2522

$I^{(2)}(\tau,\theta,\psi)$

τ	0	15	30	45	60	75	0	15	30	45	60	75
0	0,2011	0,1349	0,1078	0,1090	0,1094	0,1969	0,1272	0,0909	0,0877	0,0811	0,1162	0,1935
0,05	0,1525	0,1038	0,0833	0,0842	0,0874	0,1627	0,0981	0,0710	0,0679	0,0641	0,0923	0,1622
0,1	0,0988	0,0697	0,0570	0,0578	0,0636	0,1210	0,0665	0,0496	0,0465	0,0461	0,0661	0,1236
0,125	0,0714	0,0519	0,0439	0,0444	0,0520	0,0973	0,0500	0,0383	0,0360	0,0372	0,0527	0,1012
0,15	0,0420	0,0339	0,0304	0,0306	0,0399	0,0730	0,0326	0,0272	0,0248	0,0284	0,0388	0,0786
0,16	0,0343	0,0278	0,0244	0,0245	0,0323	0,0595	0,0263	0,0219	0,0199	0,0229	0,0314	0,0643
0,17	0,0252	0,0202	0,0184	0,0185	0,0243	0,0457	0,0198	0,0164	0,0150	0,0176	0,0238	0,0492
0,18	0,0171	0,0135	0,0124	0,0126	0,0163	0,0311	0,0134	0,0112	0,0101	0,0118	0,0161	0,0334
0,19	0,0080	0,0066	0,0063	0,0063	0,0084	0,0159	0,0066	0,0055	0,0052	0,0059	0,0082	0,0173
0,195	0,0045	0,0035	0,0032	0,0032	0,0041	0,0079	0,0035	0,0028	0,0026	0,0030	0,0042	0,0085
0,197	0,0027	0,0023	0,0019	0,0017	0,0024	0,0047	0,0019	0,0017	0,0015	0,0016	0,0025	0,0052
0,2	0	0	0	0	0	0	0	0	0	0	0	0

Table I   67

TABLE I(continued)

60						75						ζ,°
0	15	30	45	60	75	0	15	30	45	60	75	θ,° / τ
0	0	0	0	0	0	0	0	0	0	0	0	0
0,0120	0,0114	0,0147	0,0203	0,0284	0,0492	0,0092	0,0099	0,0096	0,0120	0,0190	0,0377	0,05
0,0243	0,0230	0,0300	0,0410	0,0567	0,0931	0,0196	0,0210	0,0203	0,0252	0,0394	0,0744	0,1
0,0306	0,0292	0,0376	0,0513	0,0703	0,1130	0,0252	0,0271	0,0262	0,0322	0,0501	0,0925	0,125
0,0372	0,0353	0,0455	0,0619	0,0841	0,1318	0,0313	0,0335	0,0323	0,0397	0,0612	0,1106	0,15
0,0410	0,0395	0,0502	0,0682	0,0925	0,1451	0,0346	0,0370	0,0364	0,0449	0,0687	0,1233	0,16
0,0448	0,0437	0,0555	0,0745	0,1008	0,1581	0,0380	0,0406	0,0408	0,0503	0,0764	0,1360	0,17
0,0487	0,0479	0,0607	0,0808	0,1094	0,1707	0,0414	0,0444	0,0452	0,0559	0,0841	0,1487	0,18
0,0526	0,0524	0,0656	0,0873	0,1177	0,1834	0,0449	0,0482	0,0497	0,0616	0,0919	0,1616	0,19
0,0546	0,0544	0,0684	0,0906	0,1221	0,1895	0,0467	0,0502	0,0521	0,0646	0,0959	0,1680	0,195
0,0554	0,0552	0,0694	0,0919	0,1237	0,1920	0,0475	0,0510	0,0529	0,0657	0,0976	0,1706	0,197
0,0565	0,0566	0,0709	0,0938	0,1260	0,1956	0,0485	0,0523	0,0544	0,0674	0,1000	0,1745	0,2

0	15	30	45	60	75	0	15	30	45	60	75	τ
0,0843	0,0768	0,0707	0,0939	0,1142	0,1877	0,0594	0,0506	0,0527	0,0714	0,0685	0,1796	0
0,0667	0,0606	0,0561	0,0594	0,0919	0,1588	0,0488	0,0420	0,0439	0,0591	0,0600	0,1580	0,05
0,0472	0,0424	0,0402	0,0438	0,0665	0,1215	0,0357	0,0318	0,0335	0,0442	0,0502	0,1242	0,1
0,0369	0,0330	0,0318	0,0357	0,0531	0,0994	0,0282	0,0261	0,0277	0,0357	0,0447	0,1015	0,125
0,0260	0,0229	0,0234	0,0275	0,0389	0,0766	0,0202	0,0198	0,0214	0,0264	0,0391	0,0763	0,15
0,0210	0,0184	0,0190	0,0223	0,0316	0,0677	0,0165	0,0161	0,0175	0,0216	0,0321	0,0632	0,16
0,0159	0,0139	0,0145	0,0170	0,0241	0,0481	0,0126	0,0123	0,0134	0,0165	0,0246	0,0489	0,17
0,0107	0,0094	0,0097	0,0114	0,0162	0,0330	0,0086	0,0083	0,0091	0,0112	0,0168	0,0338	0,18
0,0054	0,0048	0,0050	0,0058	0,0082	0,0169	0,0044	0,0042	0,0047	0,0057	0,0085	0,0173	0,19
0,0028	0,0025	0,0025	0,0029	0,0042	0,0084	0,0021	0,0021	0,0023	0,0029	0,0043	0,0088	0,195
0,0016	0,0014	0,0014	0,0018	0,0025	0,0051	0,0014	0,0013	0,0014	0,0017	0,0026	0,0053	0,197
0	0	0	0	0	0	0	0	0	0	0	0	0,2

0	15	30	45	60	75	0	15	30	45	60	75	τ
0	0	0	0	0	0	0	0	0	0	0	0	0
0,0120	0,0122	0,0187	0,0241	0,0360	0,0651	0,0092	0,0092	0,0100	0,0165	0,0252	0,0515	0,05
0,0243	0,0249	0,0381	0,0490	0,0717	0,1238	0,0196	0,0196	0,0213	0,0349	0,0525	0,1026	0,1
0,0306	0,0314	0,0479	0,0614	0,0894	0,1510	0,0252	0,0252	0,0274	0,0449	0,0670	0,1280	0,125
0,0372	0,0379	0,0580	0,0740	0,1072	0,1767	0,0313	0,0312	0,0338	0,0555	0,0822	0,1538	0,15
0,0410	0,0423	0,0640	0,0819	0,1167	0,1941	0,0346	0,0348	0,0389	0,0621	0,0922	0,1687	0,16
0,0448	0,0467	0,0701	0,0899	0,1261	0,2116	0,0380	0,0387	0,0429	0,0690	0,1022	0,1836	0,17
0,0487	0,0512	0,0762	0,0977	0,1358	0,2282	0,0414	0,0426	0,0477	0,0760	0,1125	0,1987	0,18
0,0526	0,0557	0,0821	0,1057	0,1452	0,2448	0,0449	0,0464	0,0525	0,0832	0,1229	0,2139	0,19
0,0546	0,0580	0,0854	0,1095	0,1503	0,2531	0,0467	0,0485	0,0551	0,0869	0,1283	0,2216	0,195
0,0554	0,0588	0,0867	0,1112	0,1522	0,2564	0,0475	0,0493	0,0560	0,0883	0,1304	0,2245	0,197
0,0565	0,0602	0,0866	0,1136	0,1548	0,2612	0,0485	0,0506	0,0575	0,0905	0,1337	0,2293	0,2

0	15	30	45	60	75	0	15	30	45	60	75	τ
0,0843	0,0734	0,0620	0,0810	0,1116	0,1859	0,0594	0,0476	0,0575	0,0703	0,1362	0,2427	0
0,0667	0,0578	0,0495	0,0645	0,0909	0,1594	0,0488	0,0397	0,0476	0,0588	0,1134	0,2143	0,05
0,0472	0,0405	0,0365	0,0466	0,0672	0,1259	0,0357	0,0304	0,0356	0,0448	0,0828	0,1684	0,1
0,0369	0,0315	0,0299	0,0371	0,0546	0,1055	0,0282	0,0252	0,0288	0,0369	0,0645	0,1338	0,125
0,0260	0,0221	0,0229	0,0273	0,0417	0,0851	0,0202	0,0195	0,0214	0,0283	0,0436	0,1005	0,15
0,0210	0,0178	0,0185	0,0222	0,0339	0,0699	0,0165	0,0158	0,0173	0,0230	0,0358	0,0832	0,16
0,0159	0,0134	0,0140	0,0169	0,0258	0,0536	0,0126	0,0121	0,0133	0,0177	0,0274	0,0646	0,17
0,0107	0,0091	0,0094	0,0313	0,0174	0,0368	0,0086	0,0083	0,0091	0,0119	0,0189	0,0447	0,18
0,0054	0,0046	0,0048	0,0058	0,0088	0,0189	0,0044	0,0041	0,0047	0,0061	0,0096	0,0230	0,19
0,0028	0,0023	0,0023	0,0029	0,0044	0,0094	0,0021	0,0021	0,0023	0,0031	0,0048	0,0116	0,195
0,0016	0,0014	0,0014	0,0017	0,0026	0,0056	0,0014	0,0013	0,0014	0,0018	0,0029	0,0070	0,197
0	0	0	0	0	0	0	0	0	0	0	0	0,2

τ* = 0,4; funct. VI; ψ = 0° $I^{(1)}(\tau, \theta, \psi)$

ζ,°	30						45					
θ,° \ τ	0	15	30	45	60	75	0	15	30	45	60	75
0	0	0	0	0	0	0	0	0	0	0	0	0
0,1	0,0348	0,0277	0,0333	0,0424	0,0546	0,139(	),0254	0,0258	0,0315	0,0358	0,0677	0,1505
0,2	0,0698	0,0551	0,0658	0,0824	0,0956	0,2421	),0515	0,0521	0,0631	0,0703	0,1222	0,2652
0,25	0,0873	0,0685	0,0817	0,1020	0,1251	0,283!	),06/5	0,0652	0,0788	0,0871	0,1589	0,3130
0,3	0,1046	0,0819	0,0971	0,1207	0,1454	0,3189	),0776	0,0784	0,0943	0,1035	0,1865	0,3543
0,32	0,1151	0,0906	0,1058	0,1301	0,1582	0,3325	0,0861	0,0862	0,1023	0,1132	0,1983	0,3714
0,34	0,1255	0,0996	0,1146	0,1338	0,1704	0,3455	0,0946	0,0941	0,1104	0,1230	0,2099	0,3876
0,36	0,1358	0,1083	0,1232	0,1483	0,1825	0,357/	0,1031	0,1020	0,1181	0,1326	0,2211	0,4033
0,38	0,1461	0,1172	0,1319	0,1573	0,1939	0,3690	0,1116	0,1099	0,1260	0,1420	0,2321	0,4183
0,39	0,1513	0,1216	0,1362	0,1617	0,1997	0,3746	0,1159	0,1139	0,1299	0,1468	0,2377	0,4257
0,395	0,1539	0,1238	0,1382	0,1639	0,2024	0,3773	0,118!	0,1159	0,1319	0,1490	0,2402	0,4295
0,397	0,1548	0,1246	0,1390	0,1649	0,2037	0,3784	0,1188	0,1166	0,1325	0,1499	0,2414	0,4310
0,4	0,1564	0,1260	0,1402	0,1662	0,2051	0,3800	0,1202	0,1178	0,1337	0,1514	0,2430	0,4329

$I^{(2)}(\tau, \theta, \psi)$

τ	0	15	30	45	60	75	0	15	30	45	60	75
0	0,3596	0,6731	0,9887	0,8650	0,6271	0,6069	0,2296	0,3540	0,6914	1,0982	1,0558	0,8467
0,1	0,2827	0,5237	0,7718	0,6849	0,5147	0,5384	0,1843	0,2822	0,5492	0,8797	0,8733	0,7593
0,2	0,1913	0,3451	0,4972	0,4543	0,3616	0,4127	0,1418	0,1941	0,3655	0,5815	0,6021	0,5838
0,25	0,1407	0,2395	0,3365	0,3169	0,2695	0,3376	0,0994	0,1429	0,2558	0,3988	0,4286	0,4548
0,3	0,0852	0,1236	0,1586	0,1649	0,1651	0,2332	0,0664	0,0873	0,1343	0,1900	0,2258	0,2899
0,32	0,0695	0,1004	0,1292	0,1341	0,1355	0,1944	0,0538	0,0709	0,1091	0,1558	0,1865	0,2429
0,34	0,0533	0,0772	0,0988	0,1034	0,1043	0,1519	0,0413	0,0545	0,0841	0,1197	0,1436	0,1903
0,36	0,0355	0,0509	0,0670	0,0708	0,0713	0,1055	0,0279	0,0368	0,0576	0,0819	0,0987	0,1328
9,38	0,0179	0,0259	0,0342	0,0361	0,0365	0,0449	0,0143	0,0188	0,0295	0,0419	0,0507	0,0693
0,39	0,0084	0,0124	0,0171	0,0184	0,0186	0,0281	0,0071	0,0093	0,0150	0,0211	0,0258	0,0355
0,395	0,0048	0,0066	0,0088	0,0092	0,0092	0,0140	0,0036	0,0048	0,0075	0,0106	0,0129	0,0179
0,397	0,0030	0,0044	0,0052	0,0049	0,0056	0,0084	0,0020	0,0027	0,0040	0,0063	0,0079	0,0107
0	0	0	0	0	0	0	0	0	0	0	0	0

τ*=0,4 funct. VI; ψ=45° $I^{(1)}(\tau, \theta, \psi)$

τ	0	15	30	45	60	75	0	15	30	45	60	75
0	0	0	0	0	0	0	0	0	0	0	0	0
0,1	0,0348	0,0296	0,0312	0,0437	0,0541	0,1187	0,0254	0,0250	0,0328	0,0358	0,0558	0,1344
0,2	0,0698	0,0590	0,0616	0,0854	0,0946	0,2061	0,0515	0,0503	0,0656	0,0707	0,0990	0,2365
0,25	0,0873	0,0735	0,0765	0,1057	0,1239	0,2411	0,0645	0,0630	0,0821	0,0876	0,1299	0,2788
0,3	0,1046	0,0879	0,0908	0,1252	0,1442	0,2702	0,0776	0,0755	0,0983	0,1041	0,1520	0,3153
0,32	0,1151	0,0970	0,0999	0,1346	0,1565	0,2859	0,0861	0,0837	0,1064	0,1136	0,1643	0,3292
0,34	0,1255	0,1063	0,1091	0,1386	0,1685	0,3007	0,0946	0,0918	0,1145	0,1230	0,1763	0,3423
0,36	0,1358	0,1153	0,1181	0,1532	0,1801	0,3145	0,1031	0,1000	0,1225	0,1324	0,1879	0,3549
0,38	0,1461	0,1242	0,1270	0,1623	0,1914	0,3277	0,1116	0,1081	0,1305	0,1416	0,1996	0,3669
0,39	0,1513	0,1288	0,1314	0,1667	0,1971	0,3337	0,1159	0,1122	0,1345	0,1461	0,2052	0,3728
0,395	0,1539	0,1311	0,1338	0,1689	0,1998	0,3369	0,1181	0,1142	0,1365	0,1484	0,2080	0,3758
0,397	0,1548	0,1320	0,1346	0,1698	0,2008	0,3381	0,1188	0,1150	0,1372	0,1494	0,2092	0,3770
0,4	0,1564	0,1333	0,1359	0,1712	0,2025	0,3400	0,1202	0,1163	0,1385	0,1507	0,2109	0,3785

$I^{(2)}(\tau, \theta, \psi)$

τ	0	15	30	45	60	75	0	15	30	45	60	75
0	0,3596	0,5113	0,5536	0,4775	0,4543	0,4786	0,2296	0,2989	0,3810	0,4389	0,4773	0,5383
0,1	0,2827	0,3995	0,4359	0,3812	0,3739	0,4265	0,1843	0,2389	0,3052	0,3556	0,3974	0,4844
0,2	0,1913	0,2676	0,2905	0,2613	0,2672	0,3264	0,1297	0,1659	0,2104	0,2476	0,2857	0,3819
0,25	0,1407	0,1907	0,2078	0,1926	0,2048	0,2766	0,0994	0,1239	0,1556	0,1845	0,2181	0,3103
0,3	0,0852	0,1067	0,1169	0,1180	0,1352	0,2042	0,0664	0,0785	0,0959	0,1144	0,1414	0,2216
0,32	0,0695	0,0865	0,0953	0,0958	0,1108	0,1702	0,0538	0,0637	0,0778	0,0936	0,1165	0,1853
0,34	0,0533	0,0666	0,0727	0,0837	0,0852	0,1328	0,0413	0,0489	0,0599	0,0718	0,0894	0,1451
0,36	0,0355	0,0439	0,0494	0,0502	0,0582	0,0922	0,0279	0,0331	0,0408	0,0489	0,0613	0,1010
0,38	0,0179	0,0223	0,0251	0,0257	0,0297	0,0479	0,0143	0,0168	0,0208	0,0249	0,0314	0,0526
0,39	0,0084	0,0107	0,0126	0,0130	0,0151	0,0245	0,0071	0,0083	0,0106	0,0125	0,0159	0,0269
0,395	0,0048	0,0057	0,0064	0,0065	0,0076	0,0122	0,0036	0,0043	0,0053	0,0063	0,0081	0,0135
0,397	0,0030	0,0038	0,0038	0,0034	0,0045	0,0074	0,0020	0,0025	0,0028	0,0038	0,0048	0,0082
0	0	0	0	0	0	0	0	0	0	0	0	0

Table I   69

TABLE I (continued)

		60						75				θ, °/τ
0	15	30	45	60	75	0	15	30	45	60	75	
0	0	0	0	0	0	0	0	0	0	0	0	0
0,0218	0,0235	0,0241	0,0405	0,0690	0,1795	0,0132	0,0128	0,0193	0,0279	0,0547	0,1506	0,1
0,0453	0,0487	0,0492	0,0821	0,1279	0,3271	0,0295	0,0284	0,0429	0,0615	0,1121	0,3062	0,2
0,0575	0,0616	0,0621	0,1034	0,1688	0,3928	0,0390	0,0373	0,0567	0,0809	0,1448	0,3909	0,25
0,0700	0,0749	0,0749	0,1248	0,2011	0,4533	0,0496	0,0472	0,0721	0,1025	0,1952	0,4829	0,3
0,0774	0,0819	0,0830	0,1340	0,2148	0,4696	0,0555	0,0535	0,0791	0,1125	0,2096	0,4999	0,32
0,0849	0,0891	0,0911	0,1435	0,2286	0,4860	0,0616	0,0600	0,0869	0,1235	0,2247	0,5187	0,34
0,0924	0,0961	0,0994	0,1529	0,2425	0,5025	0,0680	0,0669	0,0944	0,1340	0,2408	0,5397	0,36
0,1001	0,1035	0,1075	0,1623	0,2561	0,5190	0,0748	0,0741	0,1026	0,1455	0,2610	0,5629	0,38
0,1039	0,1072	0,1116	0,1672	0,2633	0,5274	0,0784	0,0779	0,1068	0,1514	0,2664	0,5755	0,39
0,1058	0,1089	0,1137	0,1696	0,2666	0,5314	0,0801	0,0798	0,1089	0,1545	0,2710	0,5819	0,395
0,1066	0,1096	0,1146	0,1706	0,2679	0,5332	0,0808	0,0807	0,1098	0,1557	0,2727	0,5845	0,397
0,1078	0,1107	0,1158	0,1721	0,2701	0,5359	0,0820	0,0818	0,1111	0,1577	0,2763	0,5886	0,4

0	15	30	45	60	75	0	15	30	45	60	75	θ, °/τ
0,1505	0,2124	0,3448	0,7239	1,2490	1,3442	0,0937	0,1139	0,1671	0,2836	0,5882	1,1263	0
0,1248	0,1753	0,2844	0,5996	1,0579	1,2364	0,0823	0,1007	0,1484	0,2549	0,5452	1,1430	0,1
0,0921	0,1271	0,2015	0,4151	0,7386	0,9487	0,0647	0,0798	0,1164	0,1986	0,4282	0,9610	0,2
0,0733	0,0986	0,1511	0,2971	0,5206	0,7109	0,0532	0,0658	0,0942	0,1567	0,3307	0,7460	0,25
0,0528	0,0667	0,0940	0,1588	0,2551	0,3894	0,0394	0,0491	0,0668	0,1032	0,1982	0,4086	0,3
0,0432	0,0547	0,0773	0,1310	0,2113	0,3281	0,0327	0,0407	0,0556	0,0863	0,1670	0,3511	0,32
0,0331	0,0420	0,0592	0,1008	0,1642	0,2592	0,0254	0,0317	0,0434	0,0675	0,1319	0,2828	0,34
0,0224	0,0286	0,0406	0,0693	0,1133	0,1821	0,0175	0,0220	0,0301	0,0470	0,0927	0,2025	0,36
0,0115	0,0145	0,0208	0,0354	0,0586	0,0961	0,0090	0,0113	0,0156	0,0245	0,0488	0,1086	0,38
0,0058	0,0073	0,0105	0,0182	0,0299	0,0491	0,0046	0,0058	0,0080	0,0125	0,0250	0,0562	0,39
0,0029	0,0037	0,0052	0,0091	0,0150	0,0248	0,0023	0,0029	0,0040	0,0064	0,0126	0,0287	0,395
0,0017	0,0022	0,0031	0,0054	0,0090	0,0149	0,0014	0,0018	0,0024	0,0037	0,0076	0,0173	0,397
0	0	0	0	0	0	0	0	0	0	0	0	

0	15	30	45	60	75	0	15	30	45	60	75	θ, °/τ
0	0	0	0	0	0	0	0	0	0	0	0	0
0,0218	0,0243	0,0241	0,0329	0,0598	0,1252	0,0132	0,0125	0,0167	0,0248	0,0390	0,0855	0,1
0,0453	0,0503	0,0491	0,0665	0,1106	0,2264	0,0295	0,0277	0,0367	0,0544	0,0774	0,1690	0,2
0,0575	0,0637	0,0620	0,0834	0,1464	0,2705	0,0390	0,0364	0,0482	0,0714	0,1077	0,2120	0,25
0,0700	0,0777	0,0750	0,1002	0,1742	0,3102	0,0496	0,0459	0,0610	0,0903	0,1395	0,2569	0,3
0,0774	0,0847	0,0829	0,1096	0,1858	0,3271	0,0555	0,0522	0,0680	0,0987	0,1467	0,2757	0,32
0,0849	0,0919	0,0908	0,1191	0,1972	0,3436	0,0616	0,0586	0,0753	0,1076	0,1594	0,2952	0,34
0,0924	0,0989	0,0987	0,1285	0,2088	0,3597	0,0680	0,0655	0,0829	0,1169	0,1729	0,3156	0,36
0,1001	0,1063	0,1068	0,1380	0,2201	0,3756	0,0748	0,0726	0,0909	0,1265	0,1901	0,3368	0,38
0,1039	0,1101	0,1108	0,1429	0,2261	0,3836	0,0784	0,0763	0,0949	0,1315	0,1940	0,3479	0,39
0,1058	0,1118	0,1127	0,1452	0,2288	0,3874	0,0801	0,0782	0,0971	0,1340	0,1977	0,3535	0,395
0,1066	0,1125	0,1135	0,1462	0,2299	0,3890	0,0808	0,0790	0,0979	0,1350	0,1991	0,3558	0,397
0,1078	0,1137	0,1149	0,1476	0,2317	0,3913	0,0820	0,0801	0,0992	0,1367	0,2021	0,3593	0,4

0	15	30	45	60	75	0	15	30	45	60	75	θ, °/τ
0,1505	0,1889	0,2495	0,3268	0,4178	0,5322	0,0937	0,1072	0,1340	0,1840	0,2909	0,3350	0
0,1248	0,1564	0,2067	0,2733	0,3579	0,4928	0,0823	0,0946	0,1193	0,1658	0,2683	0,3405	0,1
0,0921	0,1144	0,1493	0,1978	0,2650	0,3969	0,0647	0,0747	0,0950	0,1321	0,2140	0,3053	0,2
0,0733	0,0899	0,1151	0,1518	0,2057	0,3229	0,0532	0,0615	0,0784	0,1082	0,1710	0,2653	0,25
0,0528	0,0627	0,0768	0,0991	0,1360	0,2268	0,0394	0,0457	0,0583	0,0783	0,1143	0,2044	0,3
0,0432	0,0513	0,0629	0,0816	0,1123	0,1906	0,0327	0,0379	0,0485	0,0653	0,0959	0,1749	0,32
0,0331	0,0394	0,0483	0,0625	0,0870	0,1500	0,0254	0,0295	0,0378	0,0510	0,0755	0,1402	0,34
0,0224	0,0268	0,0330	0,0429	0,0599	0,1051	0,0175	0,0204	0,0262	0,0354	0,0529	0,0999	0,36
0,0115	0,0136	0,0168	0,0220	0,0309	0,0552	0,0090	0,0106	0,0136	0,0184	0,0277	0,0533	0,38
0,0058	0,0069	0,0085	0,0111	0,0156	0,0282	0,0046	0,0053	0,0070	0,0094	0,0142	0,0275	0,39
0,0029	0,0035	0,0043	0,0057	0,0079	0,0142	0,0023	0,0027	0,0035	0,0048	0,0072	0,0140	0,395
0,0017	0,0021	0,0025	0,0034	0,0048	0,0083	0,0014	0,0016	0,0021	0,0028	0,0043	0,0085	0,397
0	0	0	0	0	0	0	0	0	0	0	0	

$\tau^* = 0,4$: funct.VI; $\psi = 90°$ $I^{(1)}(\tau, \theta, \psi)$

ς.°	30						45					
θ.° \ τ	0	15	30	45	60	75	0	15	30	45	60	75
0	0	0	0	0	0	0	0	0	0	0	0	0
0,1	0,0348	0,0350	0,0343	0,0377	0,0605	0,1081	0,0254	0,0258	0,0283	0,0378	0,0582	0,1001
0,2	0,0698	0,0701	0,0679	0,0734	0,1075	0,1882	0,0515	0,0521	0,0565	0,0748	0,1045	0,1759
0,25	0,0873	0,0875	0,0845	0,0907	0,1401	0,2206	0,0645	0,0652	0,0706	0,0928	0,1369	0,2068
0,3	0,1046	0,1049	0,1005	0,1072	0,1635	0,2475	0,0776	0,0782	0,0845	0,1105	0,1607	0,2330
0,32	0,1151	0,1151	0,1106	0,1182	0,1760	0,2634	0,0861	0,0867	0,0936	0,1207	0,1724	0,2493
0,34	0,1255	0,1253	0,1207	0,1237	0,1880	0,2783	0,0946	0,0955	0,1029	0,1308	0,1840	0,2647
0,36	0,1358	0,1354	0,1306	0,1399	0,1998	0,2922	0,1031	0,1040	0,1121	0,1407	0,196?	0,2795
0,38	0,1461	0,1455	0,1404	0,1505	0,2110	0,3053	0,1116	0,1127	0,1211	0,1506	0,2061	0,2932
0,39	0,1513	0,1505	0,1452	0,1556	0,2167	0,3116	0,1159	0,1171	0,1257	0,1557	0,2116	0,3000
0,395	0,1539	0,1531	0,1477	0,1581	0,2195	0,3147	0,1181	0,1191	0,1279	0,1580	0,2143	0,3034
0,397	0,1548	0,1541	0,1487	0,1592	0,2206	0,3159	0,1188	0,1200	0,1290	0,1591	0,2154	0,3047
0,4	0,1564	0,1557	0,1501	0,1608	0,2222	0,3177	0,1202	0,1214	0,1303	0,1605	0,2170	0,3066

$I^{(2)}(\tau, \theta, \psi)$

τ	0	15	30	45	60	75	0	15	30	45	60	75
0	0,3596	0,3363	0,3009	0,2778	0,2940	0,3671	0,2296	0,2269	0,2221	0,2291	0,2705	0,3416
0,1	0,2827	0,2644	0,2388	0,2246	0,2436	0,3274	0,1843	0,1824	0,1798	0,1873	0,2262	0,3086
0,2	0,1913	0,1815	0,1651	0,1596	0,1784	0,2487	0,1297	0,1293	0,1286	0,1359	0,1665	0,2472
0,25	0,1407	0,1341	0,1245	0,1240	0,1414	0,2134	0,0994	0,0993	0,1002	0,1073	0,1320	0,2058
0,3	0,0852	0,0832	0,0810	0,0861	0,1009	0,1588	0,0664	0,0672	0,0700	0,0765	0,0938	0,1555
0,32	0,0695	0,0675	0,0658	0,0699	0,0826	0,1321	0,0538	0,0546	0,0566	0,0624	0,0770	0,1297
0,34	0,0533	0,0518	0,0501	0,0536	0,0633	0,1029	0,0413	0,0418	0,0436	0,0477	0,0589	0,1012
0,36	0,0355	0,0342	0,0339	0,0365	0,0432	0,0713	0,0279	0,0282	0,0296	0,0324	0,0402	0,0703
0,38	0,0179	0,0173	0,0173	0,0187	0,0221	0,0370	0,0143	0,0143	0,0151	0,0165	0,0205	0,0365
0,39	0,0084	0,0083	0,0087	0,0095	0,0112	0,0189	0,0071	0,0071	0,0077	0,0083	0,0104	0,0186
0,395	0,0048	0,0044	0,0045	0,0047	0,0056	0,0094	0,0036	0,0037	0,0038	0,0042	0,0053	0,0094
0,397	0,0030	0,0029	0,0026	0,0025	0,0034	0,0057	0,0020	0,0021	0,0021	0,0025	0,0032	0,0057
0,4	0	0	0	0	0	0	0	0	0	0	0	0

$\tau^* = 0,4$; funct.VI; $\psi = 135°$ $I^{(1)}(\tau, \theta, \psi)$

τ	0	15	30	45	60	75	0	15	30	45	60	75
0	0	0	0	0	0	0	0	0	0	0	0	0
0,1	0,0348	0,0380	0,0417	0,0499	0,0602	0,1008	0,0254	0,0321	0,0374	0,0460	0,0607	0,0942
0,2	0,0698	0,0761	0,0826	0,0978	0,1074	0,1759	0,0515	0,0650	0,0752	0,0915	0,1097	0,1667
0,25	0,0873	0,0951	0,1030	0,1212	0,1399	0,2063	0,0645	0,0816	0,0943	0,1140	0,1434	0,1967
0,3	0,1046	0,1139	0,1227	0,1438	0,1636	0,2320	0,0776	0,0982	0,1132	0,1360	0,1690	0,2225
0,32	0,1151	0,1255	0,1353	0,1575	0,1785	0,2520	0,0861	0,1079	0,1246	0,1494	0,1844	0,2440
0,34	0,1255	0,1371	0,1482	0,1655	0,1933	0,2711	0,0946	0,1177	0,1362	0,1627	0,1995	0,2643
0,36	0,1358	0,1486	0,1608	0,1845	0,2076	0,2888	0,1031	0,1274	0,1476	0,1758	0,2146	0,2836
0,38	0,1461	0,1602	0,1734	0,1977	0,2214	0,3057	0,1116	0,1370	0,1591	0,1888	0,2292	0,3019
0,39	0,1513	0,1660	0,1796	0,2043	0,2284	0,3136	0,1159	0,1420	0,1649	0,1955	0,2366	0,3109
0,395	0,1539	0,1689	0,1826	0,2076	0,2318	0,3176	0,1181	0,1443	0,1676	0,1987	0,2402	0,3153
0,397	0,1548	0,1699	0,1840	0,2089	0,2332	0,3191	0,1188	0,1454	0,1689	0,2000	0,2416	0,3171
0,4	0,1564	0,1717	0,1859	0,2108	0,2351	0,3214	0,1202	0,1468	0,1706	0,2020	0,2440	0,3196

$I^{(2)}(\tau, \theta, \psi)$

τ	0	15	30	45	60	75	0	15	30	45	60	75
0	0,3596	0,2672	0,2189	0,2143	0,2261	0,2854	0,2296	0,1809	0,1725	0,1711	0,1872	0,2966
0,1	0,2827	0,2111	0,1757	0,1737	0,1899	0,2600	0,1843	0,1466	0,1399	0,1417	0,1613	0,2738
0,2	0,1913	0,1472	0,1251	0,1247	0,1427	0,2049	0,1297	0,1057	0,1007	0,1047	0,1257	0,2207
0,25	0,1407	0,1112	0,0974	0,0975	0,1160	0,1852	0,0994	0,0832	0,0788	0,0841	0,1051	0,1915
0,3	0,0852	0,0726	0,0682	0,0685	0,0865	0,1494	0,0664	0,0593	0,0556	0,0617	0,0827	0,1486
0,32	0,0695	0,0588	0,0553	0,0555	0,0707	0,1242	0,0538	0,0481	0,0449	0,0504	0,0679	0,1240
0,34	0,0533	0,0452	0,0422	0,0426	0,0542	0,0968	0,0413	0,0368	0,0345	0,0384	0,0520	0,0968
0,36	0,0355	0,0296	0,0286	0,0289	0,0369	0,0669	0,0279	0,0248	0,0234	0,0261	0,0355	0,0672
0,38	0,0179	0,0151	0,0144	0,0147	0,0187	0,0348	0,0143	0,0126	0,0119	0,0132	0,0180	0,0349
0,39	0,0084	0,0072	0,0072	0,0075	0,0095	0,0177	0,0071	0,0061	0,0061	0,0066	0,0091	0,0178
0,395	0,0048	0,0039	0,0036	0,0037	0,0047	0,0088	0,0036	0,0032	0,0030	0,0033	0,0045	0,0089
0,397	0,0030	0,0025	0,0020	0,0028	0,0029	0,0054	0,0020	0,0019	0,0015	0,0020	0,0028	0,0054
0,4	0	0	0	0	0	0	0	0	0	0	0	0

Table I 71

TABLE I (continued)

60						75						θ,°
0	15	30	45	60	75	0	15	30	45	60	75	τ
0	0	0	0	0	0	0	0	0	0	0	0	0
0,0218	0,0231	0,0269	0,0341	0,0467	0,0808	0,0132	0,0137	0,0154	0,0187	0,0264	0,0492	0,1
0,0453	0,0478	0,0553	0,0691	0,0856	0,1459	0,0295	0,0306	0,0337	0,0404	0,0504	0,0950	0,2
0,0575	0,0605	0,0701	0,0871	0,1142	0,1739	0,0390	0,0404	0,0443	0,0526	0,0707	0,1173	0,25
0,0700	0,0737	0,0851	0,1052	0,1358	0,1990	0,0496	0,0512	0,0560	0,0660	0,0872	0,1398	0,3
0,0774	0,0812	0,0930	0,1141	0,1472	0,2162	0,0555	0,0573	0,0626	0,0739	0,0978	0,1564	0,32
0,0849	0,0890	0,1013	0,1233	0,1585	0,2328	0,0616	0,0636	0,0696	0,0822	0,1087	0,1731	0,34
0,0924	0,0966	0,1094	0,1324	0,1698	0,2487	0,0680	0,0702	0,0770	0,0908	0,1200	0,1899	0,36
0,1001	0,1044	0,1176	0,1416	0,1809	0,2643	0,0748	0,0772	0,0846	0,0998	0,1349	0,2071	0,38
0,1039	0,1084	0,1218	0,1462	0,1866	0,2717	0,0784	0,0809	0,0884	0,1044	0,1376	0,2159	0,39
0,1058	0,1102	0,1239	0,1484	0,1893	0,2755	0,0801	0,0826	0,0905	0,1068	0,1406	0,2203	0,395
0,1066	0,1110	0,1248	0,1494	0,1904	2771	0,0808	0,0835	0,0913	0,1077	0,1418	0,2221	0,397
0,1078	0,1122	0,1260	0,1508	0,1920	0,2792	0,0820	0,0845	0,0926	0,1092	0,1444	0,2248	0,4
0,1505	0,1528	0,1615	0,1820	0,2197	0,2739	0,0937	0,0957	0,1032	0,1173	0,1417	0,1805	0
0,1248	0,1269	0,1347	0,1535	0,1908	0,2590	0,0823	0,0843	0,0915	0,1056	0,1321	0,1852	0,1
0,0921	0,0938	0,0998	0,1145	0,1462	0,2190	0,0647	0,0665	0,0726	0,0853	0,1107	0,1717	0,2
0,0733	0,0748	0,0795	0,0913	0,1183	0,1882	0,0532	0,0547	0,0600	0,0713	0,0945	0,1553	0,25
0,0528	0,0537	0,0571	0,0654	0,0860	0,1483	0,0394	0,0406	0,0448	0,0541	0,0739	0,1300	0,3
0,0432	0,0439	0,0467	0,0537	0,0707	0,1243	0,0327	0,0336	0,0372	0,0450	0,0618	0,1108	0,32
0,0331	9,0337	0,0357	0,0411	0,0546	0,0976	0,0254	0,0262	0,0289	0,0350	0,0485	0,0884	0,34
0,0224	0,0229	0,0244	0,0281	0,0374	0,0681	0,0175	0,0181	0,0200	0,0242	0,0338	0,0628	0,36
0,0115	0,0116	0,0124	0,0143	0,0191	0,0356	0,0090	0,0093	0,0103	0,0126	0,0176	0,0333	0,38
0,0058	0,0059	0,0063	0,0072	0,0097	0,0182	0,0046	0,0047	0,0053	0,0064	0,0090	0,0171	0,39
0,0029	0,0029	0,0032	0,0037	0,0048	0,0092	0,0023	0,0024	0,0026	0,0032	0,0046	0,0087	0,395
0,0017	0,0018	0,0019	0,0022	0,0030	0,0055	0,0014	0,0014	0,0016	0,0019	0,0028	0,0053	0,397
0	0	0	0	0	0	0	0	0	0	0	0	0,4
0	0	0	0	0	0	0	0	0	0	0	0	0
0,0218	0,0213	0,0268	0,0360	0,0505	0,0861	0,0132	0,0142	0,0146	0,0183	0,0281	0,0532	0,1
0,0453	0,0441	0,0550	0,0733	0,0932	0,1566	0,0295	0,0316	0,0319	0,0393	0,0542	0,1037	0,2
0,0575	0,0558	0,0697	0,0922	0,1241	0,1873	0,0390	0,0416	0,0417	0,0511	0,0759	0,1288	0,25
0,0700	0,0677	0,0846	0,1116	0,1479	0,2153	0,0496	0,0528	0,0525	0,0639	0,0940	0,1546	0,3
0,0774	0,0758	0,0941	0,1233	0,1629	0,2368	0,0555	0,0591	0,0599	0,0732	0,1067	0,1746	0,32
0,0849	0,0842	0,1035	0,1350	0,1778	0,2574	0,0616	0,0658	0,0678	0,0829	0,1197	0,1949	0,34
0,0924	0,0924	0,1132	0,1469	0,1926	0,2776	0,0680	0,0727	0,0760	0,0930	0,1334	0,2154	0,36
0,1001	0,1008	0,1230	0,1588	0,2075	0,2970	0,0748	0,0800	0,0846	0,1036	0,1508	0,2363	0 38
0,1039	0,1052	0,1280	0,1646	0,2149	0,3065	0,0784	0,0838	0,0889	0,1090	0,1547	0,2470	0,39
0,1058	0,1071	0,1303	0,1676	0,2186	0,3111	0,0801	0,0857	0,0912	0,1118	0,1584	0,2524	0,395
0,1066	0,1080	0,1313	0,1688	0,2200	0,3131	0,0808	0,0865	0,0921	0,1129	0,1598	0,2546	0,397
0,1078	0,1093	0,1328	0,1707	0,2223	0,3158	0,0820	0,0876	0,0934	0,1146	0,1628	0,2579	0,4
0,1505	0,1376	0,1277	0,1317	0,1853	0,2582	0,0937	0,0824	0,0853	0,1079	0,1058	0,2033	0
0,1248	0,1140	0,1069	0,1129	0,1628	0,2478	0,0823	0,0731	0,0763	0,0977	0,1014	0,2100	0,1
0,0921	0,0838	0,0804	0,0878	0,1279	0,2137	0,0647	0,0588	0,0621	0,0797	0,0911	0,1938	0,2
0,0733	0,0663	0,0654	0,0733	0,1057	0,1865	0,0532	0,0496	0,0529	0,0671	0,0838	0,1727	0,25
0,0528	0,0472	0,0487	0,0576	0,0802	0,1507	0,0394	0,0388	0,0420	0,0516	0,0751	0,1394	0,3
0,0432	0,0385	0,0399	0,0473	0,0660	0,1263	0,0327	0,0321	0,0349	0,0429	0,0628	0,1189	0,32
0,0331	0,0295	0,0305	0,0361	0,0509	0,0992	0,0254	0,0250	0,0270	0,0334	0,0493	0,0950	0,34
0,0224	0,0201	0,0208	0,0246	0,0349	0,0694	0,0175	0,0172	0,0187	0,0231	0,0344	0,0675	0,36
0,0115	0,0101	0,0105	0,0125	0,0178	0,0362	0,0090	0,0089	0,0097	0,0119	0,0179	0,0358	0,38
0,0058	0,0051	0,0052	0,0063	0,0090	0,0184	0,0046	0,0045	0,0049	0,0060	0,0091	0,0185	0,39
0,0029	0,0026	0,0027	0,0033	0,0045	0,0093	0,0023	0,0023	0,0024	0,0031	0,0046	0,0094	0,395
0,0017	0,0016	0,0016	0,0019	0,0027	0,0056	0,0014	0,0013	0,0014	0,0018	0,0048	0,0057	0,397
0	0	0	0	0	0	0	0	0	0	0	0	0,4

$\tau^* = 0,4$; funct. VI; $\psi = 180°$ $I^{(1)}(\tau, \theta, \psi)$

$\zeta,°$	30						45					
$\theta,°$ τ	0	15	30	45	60	75	0	15	30	45	60	75
0	0	0	0	0	0	0	0	0	0	0	0	0
0,1	0,0348	0,0389	0,0443	0,0523	0,0722	0,1039	0,0254	0,0346	0,0391	0,0509	0,0696	0,1191
0,2	0,0698	0,0866	0,1181	0,1388	0,1307	0,1817	0,0515	0,0699	0,0766	0,1013	0,1276	0,2130
0,25	0,0873	0,0971	0,1284	0,1272	0,1687	0,2136	0,0645	0,0879	0,0987	0,1264	0,1658	0,2529
0,3	0,1046	0,1164	0,1309	0,1509	0,1978	0,2407	0,0776	0,1060	0,1185	0,1512	0,1958	0,2884
0,32	0,1151	0,1282	0,1426	0,1657	0,2154	0,2626	0,0861	0,1166	0,1311	0,1646	0,2147	0,3141
0,34	0,1255	0,1396	0,1540	0,1751	0,2328	0,2832	0,0946	0,1271	0,1436	0,1778	0,2334	0,3385
0,36	0,1358	0,1637	0,1657	0,1955	0,2493	0,3026	0,1031	0,1376	0,1563	0,1913	0,2520	0,3615
0,38	0,1461	0,1627	0,1770	0,2099	0,2656	0,3210	0,1116	0,1482	0,1688	0,2044	0,2699	0,3841
0,39	0,1513	0,1683	0,1827	0,2171	0,2738	0,3300	0,1159	0,1535	0,1751	0,2109	0,2790	0,3949
0,395	0,1539	0,1713	0,1855	0,2209	0,2779	0,3340	0,1181	0,1560	0,1780	0,2143	0,2836	0,4004
0,397	0,1548	0,1725	0,1867	0,2223	0,2796	0,3355	0,1188	0,1572	0,1795	0,2157	0,2854	0,4023
0,4	0,1564	0,1742	0,1885	0,2244	0,2818	0,3382	0,1202	0,1587	0,1813	0,2176	0,2882	0,4056

$I^{(2)}(\tau, \theta, \psi)$

τ	0	15	30	45	60	75	0	15	30	45	60	75
0	0,3596	0,2466	0,1993	0,1983	0,1974	0,3037	0,2296	0,1688	0,1603	0,1471	0,1960	0,2800
0,1	0,2827	0,1957	0,1604	0,1608	0,1683	0,2772	0,1843	0,1369	0,1304	0,1237	0,1695	0,2636
0,2	0,1913	0,1374	0,1149	0,1165	0,1300	0,2171	0,1297	0,0992	0,0942	0,0944	0,1313	0,2242
0,25	0,1407	0,1046	0,0901	0,0916	0,1084	0,1922	0,0994	0,0781	0,0740	0,0779	0,1083	0,1951
0,3	0,0852	0,0698	0,0636	0,0648	0,0847	0,1496	0,0664	0,0560	0,0524	0,0604	0,0822	0,1579
0,32	0,0695	0,0565	0,0517	0,0525	0,0692	0,1244	0,0538	0,0455	0,0425	0,0492	0,0674	0,1319
0,34	0,0533	0,0433	0,0394	0,0403	0,0531	0,0969	0,0413	0,0348	0,0326	0,0376	0,0517	0,1030
0,36	0,0355	0,0286	0,0266	0,0274	0,0362	0,0674	0,0279	0,0235	0,0221	0,0255	0,0353	0,0717
0,38	0,0179	0,0146	0,0134	0,0138	0,0183	0,0349	0,0143	0,0119	0,0112	0,0129	0,0180	0,0374
0,39	0,0084	0,0069	0,0068	0,0071	0,0094	0,0177	0,0071	0,0059	0,0057	0,0065	0,0092	0,0190
0,395	0,0048	0,0037	0,0035	0,0035	0,0047	0,0089	0,0036	0,0031	0,0029	0,0032	0,0046	0,0097
0,397	0,0030	0,0024	0,0020	0,0019	0,0028	0,0054	0,0020	0,0017	0,0015	0,0020	0,0028	0,0059
0,4	0	0	0	0	0	0	0	0	0	0	0	0

$\tau^* = 0,4$ funct. VII; $\psi = 0°$ $I^{(1)}(\tau, \theta, \psi)$

τ	0	15	30	45	60	75	0	15	30	45	60	75
0	0	0	0	0	0	0	0	0	0	0	0	0
0,1	0,0200	0,0164	0,0234	0,0309	0,0438	0,1239	0,0175	0,0145	0,0227	0,0295	0,0549	0,1522
0,2	0,0401	0,0324	0,0461	0,0600	0,0754	0,2118	0,0353	0,0289	0,0453	0,0578	0,0971	0,2641
0,25	0,0499	0,0405	0,0573	0,0739	0,0987	0,2465	0,0443	0,0362	0,0566	0,0614	0,1264	0,3092
0,3	0,0597	0,0482	0,0681	0,0872	0,1139	0,2741	0,0531	0,0432	0,0676	0,0846	0,1472	0,3468
0,32	0,0704	0,0571	0,0771	0,0968	0,1268	0,2896	0,0616	0,0513	0,0755	0,0941	0,1596	0,3633
0,34	0,0811	0,0662	0,0857	0,1062	0,1392	0,3044	0,0702	0,0595	0,0835	0,1039	0,1719	0,3795
0,36	0,0919	0,0752	0,0944	0,1155	0,1516	0,3180	0,0787	0,0676	0,0916	0,1134	0,1839	0,3948
0,38	0,1026	0,0841	0,1032	0,1246	0,1636	0,3312	0,0873	0,0757	0,0996	0,1230	0,1957	0,4097
0,39	0,1079	0,0885	0,1075	0,1288	0,1693	0,3375	0,0917	0,0798	0,1036	0,1276	0,2017	0,4168
0,395	0,1105	0,0909	0,1097	0,1316	0,1720	0,3405	0,0936	0,0819	0,1055	0,1299	0,2045	0,4205
0,397	0,1116	0,0917	0,1105	0,1323	0,1733	0,3418	0,0945	0,0827	0,1064	0,1308	0,2056	0,4219
0,4	0,1133	0,0931	0,1117	0,1338	0,1747	0,3437	0,0959	0,0839	0,1075	0,1322	0,2074	0,4241

$I^{(2)}(\tau, \theta, \psi)$

τ	0	15	30	45	60	75	0	15	30	45	60	75
0	0,4337	0,8946	1,3423	1,1469	0,7744	0,7059	0,2587	0,4312	0,9094	1,4931	1,4101	1,0449
0,1	0,3324	0,6810	1,0279	0,8890	0,6176	0,6065	0,2027	0,3298	0,7077	1,1739	1,1411	0,9105
0,2	0,2157	0,4295	0,6354	0,5632	0,4133	0,4494	0,1378	0,2211	0,4510	0,7445	0,7519	0,6674
0,25	0,1521	0,2822	0,4073	0,3721	0,2940	0,3476	0,1025	0,1558	0,2995	0,4839	0,5069	0,4968
0,3	0,0832	0,1214	0,1559	0,1620	0,1610	0,2255	0,0646	0,0852	0,1321	0,1873	0,2221	0,2831
0,32	0,0679	0,0986	0,1272	0,1318	0,1321	0,1882	0,0524	0,0693	0,1075	0,1537	0,1832	0,2373
0,34	0,0521	0,0759	0,0974	0,1016	0,1018	0,1472	0,0403	0,0533	0,0826	0,1181	0,1412	0,1861
0,36	0,0348	0,0500	0,0661	0,0695	0,0694	0,1024	0,0272	0,0360	0,0567	0,0806	0,0972	0,1299
0,38	0,0175	0,0255	0,0337	0,0355	0,0357	0,0533	0,0139	0,0184	0,0290	0,0413	0,0499	0,0677
0,39	0,0082	0,0122	0,0169	0,0181	0,0181	0,0273	0,0097	0,0092	0,0148	0,0208	0,0254	0,0347
0,395	0,0047	0,0065	0,0087	0,0091	0,0090	0,0136	0,0036	0,0047	0,0074	0,0105	0,0144	0,0175
0,397	0,0028	0,0044	0,0051	0,0048	0,0054	0,0081	0,0021	0,0027	0,0040	0,0063	0,0079	0,0105
0,4	0	0	0	0	0	0	0	0	0	0	0	0

Table I 73

		60						75				θ,° / τ
0	15	30	45	60	75	0	15	30	45	60	75	
0	0	0	0	0	0	0	0	0	0	0	0	0
0,0218	0,0226	0,0324	0,0418	0,0616	0,1092	0,0132	0,0136	0,0151	0,0230	0,0350	0,0688	0,1
0,0453	0,0468	0,0667	0,0851	0,1152	0,1999	0,0295	0,0301	0,0330	0,0504	0.0694	0,1361	0,2
0,0575	0,0591	0,0847	0,1072	0,1524	0,2402	0,0390	0,0396	0,0432	0,0662	0,0966	0,1707	0,25
0,0700	0,0718	0,1031	0,1299	0,1822	0,2774	0,0496	0,0501	0,0545	0,0836	0,1205	0.2070	0,3
0,0774	0,0801	0,1144	0,1442	0,1988	0,3052	0,0555	0,0566	0,0624	0,0951	0,1370	0,2297	0,32
0,0849	0,0887	0,1255	0,1583	0,2155	0,3321	0,0616	0,0635	0.0708	0,1071	0,1540	0,2529	0,34
0,0924	0,0973	0,1369	0,1730	0,2319	0.3583	0,0680	0,0707	0,0796	0,1197	0,1720	0,2767	0,36
0,1001	0,1060	0,1485	0,1876	0,2484	0,3839	0,0748	0,0782	0,0887	0,1329	0,1939	0,3012	0,38
0,1039	0,1105	0,1542	0,1948	0,2569	0,3965	0,0784	0,0822	0,(934	0,1397	0,2002	0,3139	0,39
0,1058	0,1125	0,1571	0,1986	0,2611	0,4027	0,0801	0.0842	0,(95)	0,1433	0,2051	0,3202	0,395
0,1066	0,1134	0.1583	0,2000	0,2626	0,4053	0,0808	0,0851	0,09.8	0,1446	0,2070	0,3288	0,397
0,1078	0,1148	0,1601	0,2023	0,2651	0,4092	0,0820	0,0862	0,0983	0,1468	0,2108	0,3268	0,4

0	15	30	45	60	75	0	15	30	45	60	75	θ,° / τ
0,1505	0,1318	0,1145	0,1426	0,1856	0,2642	0,0937	0,0785	0,0908	0,1078	0,1825	0,2698	0
0,1248	0,1091	0 0967	0,1214	0,1638	0,2544	0,0823	0,0698	0,0810	0,0983	0,1703	0,2801	0,1
0,0921	0,0804	0,0741	0,0926	0,1302	0,2232	0,0647	0,0566	0,0651	0,0813	0,1398	0,2582	0,2
0,0733	0.0638	0,0614	0,0757	0,1095	0,1985	0,0532	0,0482	0,0547	0,0695	0,1155	0,2283	0,25
0,0528	0,0456	0,0477	0,0573	0,0857	0,1666	0,0394	0,0383	0,0421	0,0551	0,0834	0,1805	0,3
0,0432	0,0372	0,0390	0,0471	0,0705	0,1399	0,0327	0,0318	0,0349	0,0459	0,0698	0.1543	0,32
0,0331	0,0284	0,0298	0,0360	0,0545	0,1099	0,0254	0,0247	0,0271	0,0357	0,0548	0,1236	0,34
0,0224	0,0194	0,0203	0,0244	0,0374	0,0769	0,0175	0,0171	0,0187	0,0247	0,0383	0,0881	0,36
0,0115	0,0098	0.0103	0,0124	0,0192	0,0403	0,0090	0,0088	0,0097	0,0128	0,0200	0,0470	0,38
0,0058	0,0050	0,0053	0,0064	0,0097	0,0205	0,0046	0,0044	0.0049	0,0065	0,0102	0.0242	0,39
0,0029	0,0025	0,0026	0,0032	0,0049	0,0104	0,0023	0,0022	0,0025	0,0033	0,0052	0,0124	0,395
0,0017	0,0015	0,0015	0,0019	0,0029	0,0063	0,0014	0,0014	0,0015	0,0019	0,0032	0,0074	0,397
0	0	0	0	0	0	0	0	0	0	0	0	0,4

0	15	30	45	60	75	0	15	30	45	60	75	θ,° / τ
0	0	0	0	0	0	0	0	0	0	0	0	0
0,0145	0,0164	0,0187	0,0324	0,0679	0.2138	0,0091	0,0101	0,0160	0,0269	0,0627	0,1892	0,1
0,0300	0,0337	0,0381	0,0652	0,1241	0,3851	0,0204	0,0226	0,0358	0,0596	0,1293	0,3831	0,2
0,0381	0,0426	0,0480	0,0817	0,1631	0,4600	0,0270	0,0298	0,0473	0,0784	0,1783	0.4899	0,25
0,0463	0,0518	0,0578	0,0979	0,1931	0,5274	0,0343	0,0378	0,0603	0,0995	0,2250	0,6011	0,3
0,0538	0.0590	0,0658	0,1075	0,2066	0,5369	0,0404	0,0441	0.0676	0,1093	0,2381	0,6089	0,32
0,0613	0,0660	0,0738	0,1170	0,2199	0,5471	0,0467	0,0508	0,0749	0,1197	0,2518	0,6193	0,34
0,0690	0,0734	0,0821	0,1268	0,2336	0,5580	0,0532	0,0576	0,0829	0,1307	0,2666	0,6325	0,36
0,0768	0,0807	0,0902	0,1365	0,2472	0,5692	0,0601	0,0649	0,0912	0,1420	0.2823	0,6484	0,38
0,0807	0,0844	0,0944	0,1415	0,2540	0,5752	0,0636	0,0688	0,0955	0,1480	0,2905	0,6575	0,39
0,0826	0,0862	0,0964	0,1440	0,2573	0,5782	0,0655	0,0707	0,0977	0,1511	0,2947	0,6623	0,395
0,0835	0,0870	0,0974	0,1449	0,2588	0,5794	0,0662	0,0714	0,0985	0,1522	0,2964	0,6643	0,397
0,0847	0,0881	0,0987	0,1466	0,2608	0,5812	0,0674	0,0726	0,0999	0,1541	0,2970	0,6674	0,4

0	15	30	45	60	75	0	15	30	45	60	75	θ,° / τ
0,1468	0,2384	0.4172	0,9593	1,7091	1,8207	0,0848	0,1119	0,1856	0,3369	0,8272	1,5989	0
0,1201	0,1926	0,3360	0,7781	1,4189	1,6358	0,0741	0,0978	0,1619	0,2966	0,7495	1,6011	0,1
0,0879	0,1349	0,2283	0,5150	0,9497	1,1974	0,0590	0,0771	0,1234	0,2225	0,5588	1,2779	0,2
0,0703	0,1017	0,1643	0,3493	0,6339	0,8486	0,0496	0,0639	0,0976	0,1693	0,4043	0,9343	0,25
0,0514	0,0652	0,0924	0,1568	0,2522	0,3844	0,0388	0,0483	0,0661	0,1022	0,1970	0,4067	0,3
0,0420	0,0535	0,0759	0,1294	0,2092	0,3237	0,0322	0,0401	0,0550	0,0855	0,1660	0,3494	0,32
0,0322	0,0411	0.0581	0,0997	0,1626	0,2561	0,0249	0,0313	0,0428	0,0670	0,1312	0,2816	0,34
0,0220	0,0280	0,0399	0,0683	0,1122	0,1800	0,0172	0,0216	0,0298	0,0467	0,0922	0,2016	0,36
0.0111	0,0142	0,0205	0,0350	0,0580	0,0950	0,0090	0,0112	0,0155	0,0243	0,0486	0,1083	0,38
0,0057	0;0072	0.0104	0,0179	0,0294	0,0487	0,0045	0,0057	0,0079	0,0124	0,0248	0,0560	0,39
0,0028	0,0037	0,0052	0,0089	0,0148	0,0244	0,0023	0,0028	0,0040	0,0062	0.0124	0,0284	0,395
0,0018	0,0021	0,0031	0,0054	0,0089	0,0148	0,0014	0,0017	0,0024	0,0037	0,0075	0,0172	0,397
0	0	0	0	0	0	0	0	0	0	0	0	0,4

$\tau^* = 0,4$ funct. VII; $\psi = 45°$ $I^{(1)} (\tau, \theta, \psi)$

$\zeta,°$	30						45					
τ \ $\theta,°$	0	15	30	45	60	75	0	15	30	45	60	75
0	0	0	0	0	0	0	0	0	0	0	0	0
0,1	0,0200	0.0183	0,0219	0,0307	0,0412	0,1076	0,0175	0,0148	0,0225	0,0274	0,0465	0,1213
0,2	0,0401	0,0362	0,0433	0,0597	0,0706	0,1837	0,0353	0,0297	0,0452	0,0535	0,0810	0,2092
0,15	0,0499	0,0450	0,0538	0,0736	0,0929	0,2131	0,0443	0,0371	0,0564	0,0662	0,1065	0,2440
0,3	0,0597	0,0536	0,0639	0,0869	0,1071	0,2366	0,0531	0,0442	0,0675	0,0784	0,1237	0,2725
0.32	0,0704	0,0630	0,0732	0,0968	0,1199	0,2532	0,0616	0,0526	0,0757	0,0879	0,1362	0,2883
0,34	0,0811	0,0723	0,0823	0,1064	0,1324	0,2691	0,0702	0,0609	0,0838	0,0974	0,1486	0,3034
0,36	0,0919	0,0815	0,0913	0,1160	0,1446	0,2836	0,0787	0,0693	0,0922	0,1068	0,1605	0,3177
0,38	0,1026	0,0907	0,1002	0,1254	0,1563	0,2978	0,0873	0,0776	0,1003	0,1163	0,1723	0,3322
0,39	0,1079	0,0953	0,1046	0,1296	0,1622	0,3043	0,0917	0,0817	0,1043	0,1208	0,1781	0,3381
0,395	0,1105	0,0976	0,1070	0,1324	0,1649	0,3077	0,0936	0,0838	0,1064	0,1231	0,1810	0,3416
0,397	0,1116	0,0984	0,1078	0,1333	0,1660	0,3090	0,0945	0,0846	0,1072	0,1240	0,1821	0,3428
0,4	0,1133	0,0998	0,1092	0,1349	0,1677	0,3110	0,0959	0,0858	0,1085	0,1254	0,1839	0,3448

$I^{(2)} (\tau, \theta, \psi)$

τ	0	15	30	45	60	75	0	15	30	45	60	75
0	0,4337	0,6555	0,7054	0,5813	0,5377	0,5156	0,2587	0,3535	0,4581	0,5313	0.5735	0,6247
0,1	0.3324	0.5002	0,5419	0,4510	0,4286	0,4439	0,2027	0,2715	0.3577	0,4186	0,4625	0,5424
0,2	0,2157	0,3202	0,3451	0,2949	0,2919	0,3370	0,1378	0,1836	0,2362	0,2783	0,3164	0,4073
0,25	0.1521	0,2166	0,2344	0,2081	0,2151	0,2720	0,1025	0,1319	0,1677	0.1988	0.2318	0,3198
0,3	0.0832	0,1045	0,1144	0,1152	0,1312	0,1967	0,0646	0,0765	0,0937	0,1117	0.1377	0,2147
0,32	0,0679	0.0848	0,0933	0,0935	0,1075	0,1642	0,0524	0,0622	0,0762	0,0914	0,1133	0,1798
0,34	0,0521	0,0654	0,0713	0,0720	0,0827	0,1282	0,0403	0,0478	0,0585	0,0702	0,0871	0,1409
0,36	0,0348	0,0431	0,0485	0,0490	0,0563	0,0812	0,0272	0,0323	0,0399	0,0478	0,0598	0,0980
0,38	0,0175	0,0219	0,0246	0,0251	0,0289	0,0463	0,0139	0,0164	0,0204	0,0244	0,0306	0,0510
0,39	0,0082	0,0105	0,0123	0,0127	0,0147	0,0238	0,0097	0,0081	0,0104	0,0123	0,0156	0,0261
0.395	0,0047	0,0057	0,0063	0,0065	0,0074	0,0118	0,0036	0,0042	0,0052	0,0062	0,0091	0,0131
0,397	0,0028	0,0038	0,0037	0,0033	0,0043	0,0071	0,0021	0,0024	0,0028	0,0038	0,0048	0,0080
0	0	0	0	0	0	0	0	0	0	0	0	0

$\tau^* = 0,4$; funct. VII; $\psi = 90°$ $I^{(1)} (\tau, \theta, \psi)$

τ	0	15	30	45	60	75	0	15	30	45	60	75
0	0	0	0	0	0	0	0	0	0	0	0	0
0,1	0,0200	0,0214	0,0233	0,0263	0,0424	0,0843	0,0175	0,0178	0,0194	0,0258	0,0417	0,0811
0,2	0,0401	0,0425	0,0460	0,0510	0,0739	0,1438	0,0353	0,0358	0.0386	0,0507	0,0731	0,1394
0,25	0,0499	0,0530	0,0573	0.0630	0,0968	0,1671	0,0443	0,0448	0,0483	0,0629	0,0966	0,1622
0,3	0,0597	0.0633	0,0681	0,0741	0,1124	0,1855	0,0531	0,0537	0,0577	0,0746	0,1127	0,1804
0,32	0,0704	0,0738	0,0783	0,0854	0,1256	0,2040	0,0616	0,0623	0,0670	0,0850	0,1252	0,1989
0,34	0,0811	0,0842	0,0886	0,0965	0,1387	0,2215	0,0702	0,0711	0,0762	0,0956	0,1376	0,2163
0,36	0,0919	0,0948	0,0986	0,1073	0,1514	0,2378	0,0787	0,0797	0,0855	0,1058	0,1498	0,2330
0,38	0,1026	0,1051	0,1086	0,1181	0,1637	0,2532	0,0873	0,0883	0,0946	0,1161	0,1617	0,2486
0,39	0,1079	0,1104	0,1135	0,1234	0,1698	0,2607	0,0917	0,0927	0,0992	0,1212	0,1675	0,2562
0,395	0,1105	0,1130	0,1161	0,1260	0,1728	0,2644	0,0936	0,0949	0.1015	0,1237	0,1705	0,2600
0,397	0,1116	0,1139	0,1170	0,1272	0,1740	0,2657	0,0945	0,0957	0,1025	0,1247	0,1716	0,2614
0,4	0,1133	0,1155	0,1185	0,1288	0,1758	0,2678	0,0959	0,0971	0,1037	0,1262	0,1734	0,2637

$I^{(2)} (\tau, \theta, \psi)$

τ	0	15	30	45	60	75	0	15	30	45	60	75
0	0,4337	0,3993	0,3495	0,3069	0,2857	0,3425	0,2587	0,2539	0,2416	0,2326	0,2573	0,3276
0,1	0,3324	0,3058	0.2694	0,2399	0,2298	0,2969	0,2027	0,1993	0,1903	0,1848	0,2090	0,2864
0,2	0,2157	0,2012	0,1784	0,1641	0.1651	0,2317	0,1378	0,1363	0,1317	0,1307	0,1512	0,2248
0,25	0.1521	0,1431	0,1298	0,1243	0,1317	0,1939	0,1025	0,1018	0,1005	0,1028	0.1211	0,1890
0,3	0,0832	0,0812	0.0787	0,0833	0,0969	0,1514	0,0646	0,0654	0,0678	0,0738	0,0901	0,1486
0,32	0,0679	0,0658	0,0640	0,0676	0,0794	0,1261	0,0524	0,0531	0,0550	0,0603	0,0740	0,1241
0,34	0,0521	0,0506	0,0488	0,0520	0,0609	0,0984	0,0403	0,0407	0,0423	0,0462	0,0567	0,0969
0,36	0,0348	0.0334	0,0331	0,0354	0.0415	0,0684	0,0272	0,0275	0,0288	0,0314	0,0388	0,0674
0,38	0,0175	0,0170	0,0168	0,0182	0,0213	0.0354	0,0139	0,0139	0,0147	0,0160	0,0199	0,0350
0,39	0,0082	0,0081	0,0084	0,0092	0,0108	0.0181	0,0097	0,0070	0,0075	0,0081	0,0101	0,0178
0,395	0,0047	0,0043	0,0044	0,0046	0,0053	0,0091	0,0036	0,0036	0,0037	0,0040	0,0051	0,0089
0,397	0,0028	0,0029	0,0025	0,0024	0,0033	0,0054	0,0021	0,0021	0,0021	0,0024	0,0031	0,0055
0	0	0	0	0	0	0	0	0	0	0	0	0

Table I   75

TABLE I (continued)

60						75						θ,° / τ
0	15	30	45	60	75	0	15	30	45	60	75	
0	0	0	0	0	0	0	0	0	0	0	0	0
0,0145	0,0165	0,0174	0,0268	0,0516	0,1231	0,0091	0,0095	0,0138	0,0210	0,0366	0.0949	0,1
0,0300	0,0341	0,0356	0,0538	0,0934	0.2181	0,0204	0,0213	0,0308	0,0463	0,0721	0,1860	0,2
0,0381	0,0432	0,0447	0,0673	0,1237	0.2579	0 0270	0,0281	0,0405	0,0606	0,1005	0.2342	0,15
0,0463	0,0526	0,0538	0,0805	0,1462	0.2919	0,0343	0.0355	0,0515	0,0765	0,1249	0.2798	0,3
0,0538	0,0598	0,0618	0,0899	0,1583	0,3087	0,0404	0,0418	0,0586	0,0851	0,1373	0.2964	0,32
0,0613	0,0670	0,0698	0,0995	0,1702	0,3255	0,0467	0,0484	0,0658	0.0941	0,1501	0.3138	0,34
0,0690	0,0743	0.0779	0,1090	0,1822	0,3417	0.0532	0,0553	0,0735	0,1036	0,1636	0.3323	0,36
0,0768	0,0818	0,0860	0,1185	0,1939	0.3579	0,0601	0,0625	0,0816	0 1133	0,1776	0,3519	0,38
0,0807	0,0855	0,0901	0,1234	0,1999	0.3657	0,0636	0,0662	0,0856	0.1184	0,1849	0,3622	0,39
0,0826	0,0874	0,0921	0,1259	0,2028	0,3698	0,0655	0,0681	0.0878	0,1210	0,1886	0,3674	0,395
0,0835	0.0881	0,0929	0,1268	0,2040	0,3714	0,0662	0,0689	0,0886	0,1221	0,1901	0.3696	0,397
0,0847	0,0893	0,0942	0,1283	0,2059	0,3737	0,0674	0,0700	0,0899	0,1237	0,1923	0.3728	0,4
0,1468	0,2057	0,2873	0,3859	0,4979	0.6392	0,0848	0,1006	0,1383	0,2041	0,3020	0.4338	0
0,1201	0,1669	0,2323	0,3141	0,4136	0.5717	0,0741	0,0880	0,1213	0,1801	0,2729	0.4375	0,1
0.0879	0,1186	0,1615	0,2176	0,2924	0.4373	0,0590	0,0698	0 0950	0,1392	0,2131	0,3608	0,2
0,0703	0,0913	0,1206	0,1608	0,2182	0.3418	0,0496	0,0584	0 0779	0,1114	0,1691	0.2961	0,25
0,0514	0,0612	0.0753	0,0971	0,1332	0.2218	0,0388	0,0450	0,0575	0,0773	0,1131	0,2022	0,3
0,0420	0.0501	0 0617	0,0800	0,1102	0,1862	0.0322	0,0373	0,0479	0,0645	0,0950	0,1730	0.32
0,0322	0,0385	0,0473	0,0614	0,0854	0,1469	0,0249	0,0291	0,0373	0,0505	0,0748	0,1388	0,34
0,0220	0,0262	0,0323	0,0421	0,0588	0,1030	0,0172	0,0201	0,0259	0,0351	0,0524	0,0989	0,36
0,0111	0,0133	0,0165	0,0216	0,0303	0,0541	0,0090	0,0104	0.0135	0,0182	0,0275	0,0529	0,38
0,0057	0,0068	0,0084	0,0110	0,0152	0,0278	0,0045	0,0053	0,0069	0,0093	0,0140	0,0273	0,39
0,0028	0,0034	0,0042	0,0055	0,0077	0,0139	0,0023	0,0026	0,0035	0,0047	0,0070	0,0138	0,395
0,0018	0,0020	0,0025	0,0034	0,0047	0,0084	0,0014	0,0016	0,0021	0,0028	0,0042	0,0085	0,397
0	0	0	0	0	0	0	0	0	0	0	0	0
0	0	0	0	0	0	0	0	0	0	0	0	0
0,0145	0,0154	0,0181	0,0236	0,0345	0,0658	0,0091	0,0095	0,0109	0,0138	0,0207	0,0415	0,1
0,0300	0,0317	0,0371	0,0478	0,0621	0,1172	0,0204	0,0213	9,0241	0,0300	0,0393	0,0795	0,2
0,0381	0,0403	0,0469	0,0600	0,0832	0,1391	0,0270	0,0282	0,0315	0,0390	0,0554	0,0997	0,25
0,0463	0,0489	0,0568	0,0721	0,0986	0,1581	0,0343	0,0357	0,0398	0,0489	0,0683	0,1161	0,3
0,0538	0,0566	0,0650	0,0817	0,1107	0,1769	0,0404	0,0419	0,0467	0,0570	0,0793	0,1337	0,32
0,0613	0,0644	0,0735	0,0913	0,1228	0,1951	0,0467	0,0485	0,0539	0,0656	0,0905	0,1514	0,34
0,0690	0,0723	0,0819	0,10 6	0,1348	0,2125	0,0532	0,0553	0,0613	0,0744	0,1022	0,1692	0,36
0,0768	0,0802	0,0904	0,1103	0,1466	0,2296	0,0601	0,0623	0,0691	0,0835	0,1141	0,1873	0,38
0,0807	0,0842	0,0948	0,1152	0,1526	0,2377	0,0636	0,0660	0,0731	0,0883	0,1203	0,1965	0,39
0,0826	0,0862	0,0968	0,1175	0,1554	0,2420	0,0655	0,0679	0,0751	0,0907	0,1234	0,2012	0,395
0,0835	0,0870	0,0978	0,1185	0,1566	0,2436	0,0662	0,0687	0,0759	0,0917	0,1247	0,2031	0,397
0,0847	0,0882	0,0991	0,1199	0,1585	0,2459	0,0674	0,0698	0,0772	0,0932	0,1266	0,2058	0,4
0,1468	0,1472	0,1511	0,1653	0,1977	0,2511	0,0848	0,0869	0,0942	0,1084	0,1322	0,1682	0
0,1201	0,1206	0,1244	0,1378	0,1695	0,2341	0,0741	0,0761	0,0829	0,0969	0,1222	0,1817	0,1
0,0879	0,0887	0,0921	0,1033	0,1310	0,1986	0,0590	0,0607	0,0665	0,0787	0,1029	0,1597	0,2
0,0703	0,0712	0,0746	0,0841	0,1084	0,1740	0,0496	0,0511	0,0562	0,0670	0,0893	0,1409	0,25
0,0514	0,0524	0,0557	0,0634	0,0831	0,1431	0,0388	0,0400	0,0441	0,0531	0,0725	0,1274	0,3
0,0420	0,0429	0,0455	0,0521	0,0685	0,1200	0,0322	0,0331	0,0366	0,0442	0,0608	0,1086	0,32
0,0322	0,0328	0,0349	0,0400	0,0530	0,0943	0,0249	0,0258	0,0285	0,0345	0,0477	0,0869	0,34
0,0220	0,0223	0,0239	0,0274	0,0364	0,0660	0,0172	0,0178	0,0198	0,0239	0,0333	0,0617	0,36
0,0111	0,0113	0,0121	0,0140	0,0186	0,0346	0,0090	0,0092	0,0103	0,0124	0,0174	0,0329	0,38
0,0057	0,0057	0,0061	0,0071	0,0094	0,0177	0,0045	0,0047	0,0052	0,0063	0,0089	0,0169	0,39
0,0028	0,0029	0,0032	0,0035	0,0047	0,0089	0,0023	0,0023	0,0027	0,0032	0,0045	0,0086	0,395
0,0018	0,0017	0,0019	0,0022	0,0029	0,0054	0,0014	0,0014	0,0016	0,0019	0,0027	0,0053	0,397
0	0	0	0	0	0	0	0	0	0	0	0	0

τ*=0,4; funct.VII; ψ=135° $I^{(1)}(\tau, \theta, \psi)$

ζ°	30						54					
θ,° \ τ	0	15	30	45	60	75	0	15	30	45	60	75
0	0	0	0	0	0	0	0	0	0	0	0	0
0,1	0,0200	0,0219	0,0210	0,0289	0,0440	0,0705	0,0175	0,0223	0,0214	0,0278	0,0427	0,0671
0,2	0,0401	0,0433	0,0414	0,0563	0,0781	0,1212	0,0353	0,0450	0,0430	0,0550	0,0768	0,1172
0,25	0,0499	0,0542	0,0516	0,0697	0,1017	0,1413	0,0443	0,0566	0,0538	0,0684	0,1009	0,1376
0,3	0,0597	0,0647	0,0612	0,0824	0,1186	0,1577	0,0531	0,0680	0,0644	0,0816	0,1187	0,1547
0,32	0,0704	0,0767	0,0747	0,0971	0,1342	0,1807	0,0616	0,0777	0,0763	0,0956	0,1349	0,1789
0,34	0,0811	0,0886	0,0880	0,1116	0,1496	0,2027	0,0702	0,0876	0,0884	0,1096	0,1511	0,2015
0,36	0,0919	0,1007	0,1016	0,1258	0,1645	0,2232	0,0787	0,0976	0,1005	0,1236	0,1669	0,2234
0,38	0,1026	0,1127	0,1150	0,1399	0,1791	0,2427	0,0873	0,1074	0,1125	0,1376	0,1823	0,2438
0,39	0,1079	0,1188	0,1215	0,1474	0,1864	0,2519	0,0917	0,1123	0,1185	0,1444	0,1902	0,2538
0,395	0,1105	0,1218	0,1248	0,1504	0,1900	0,2566	0,0936	0,1148	0,1215	0,1480	0,1940	0,2588
0,397	0,1116	0,1229	0,1261	0,1518	0,1914	0,2585	0,0945	0,1158	0,1227	0,1492	0,1954	0,2608
0,4	0,1133	0,1247	0,1281	0,1540	0,1935	0,2612	0,0959	0,1172	0,1245	0,1513	0,1979	0,2637

$I^{(2)}(\tau, \theta, \psi)$

τ	0	15	30	45	60	75	0	15	30	45	60	75
0	0,4337	0,3022	0,2319	0,1955	0,2018	0,2532	0,2587	0,1804	0,1561	0,1545	0,1609	0,2430
0,1	0,3324	0,2330	0,1817	0,1558	0,1669	0,2271	0,2027	0,1475	0,1250	0,1264	0,1375	0,2234
0,2	0,2157	0,1566	0,1258	0,1117	0,1261	0,1892	0,1378	0,1020	0,0904	0,0940	0,1092	0,1895
0,25	0,1521	0,1149	0,0966	0,0891	0,1048	0,1669	0,1025	0,0802	0,0722	0,0771	0,0945	0,1678
0,3	0,0832	0,0709	0,0663	0,0658	0,0825	0,1417	0,0646	0,0575	0,0535	0,0592	0,0790	0,1417
0,32	0,0679	0,0573	0,0539	0,0533	0,0676	0,1181	0,0524	0,0467	0,0435	0,0484	0,0652	0,1183
0,34	0,0521	0,0441	0,0411	0,0410	0,0518	0,0921	0,0403	0,0358	0,0334	0,0369	0,0498	0,0925
0,36	0,0348	0,0290	0,0278	0,0280	0,0354	0,0639	0,0272	0,0241	0,0228	0,0252	0,0342	0,0643
0,38	0,0175	0,0147	0,0141	0,0142	0,0181	0,0333	0,0139	0,0122	0,0117	0,0128	0,0174	0,0335
0,39	0,0082	0,0070	0,0070	0,0072	0,0091	0,0170	0,0097	0,0061	0,0058	0,0064	0,0089	0,0171
0,395	0,0047	0,0038	0,0036	0,0037	0,0045	0,0086	0,0036	0,0032	0,0030	0,0032	0,0032	0,0085
0,397	0,0028	0,0025	0,0021	0,0019	0,0027	0,0051	0,0021	0,0019	0,0016	0,0019	0,0027	0,0052
0,0	0	0	0	0	0	0	0	0	0	0	0	0

τ*=0,4; funct.VII; ψ=180° $I^{(1)}(\tau, \theta, \psi)$

τ	0	15	30	45	60	75	0	15	30	45	60	75
0,1	0,0200	0,0236	0,0269	0,0276	0,0452	0,0711	0,0175	0,0228	0,0206	0,0329	0,0416	0,0691
0,2	0,0401	0,0471	6,0535	0,0540	0,0808	0,1232	0,0353	0,0461	0,0415	0,0653	0,0751	0,1122
0,25	0,0499	0,0588	0,0666	0,0666	0,1049	0,1442	0,0443	0,0578	0,0520	0,0815	0,0987	0,1446
0,3	0,0597	0,0703	0,0795	0,0790	0,1228	0,1617	0,0531	0,0697	0,0621	0,0974	0,1163	0,1642
0,32	0,0704	0,0823	0,0918	0,0951	0,1421	0,1866	0,0616	0,0803	0,0753	0,1114	0,1372	0,1963
0,34	0,0811	0,0942	0,1036	0,1110	0,1611	0,2103	0,0702	0,0913	0,0885	0,1255	0,1579	0,2267
0,36	0,0919	0,1063	0,1159	0,1271	0,1795	0,2328	0,0787	0,1021	0,1020	0,1396	0,1781	0,2556
0,38	0,1026	0,1180	0,1278	0,1426	0,1976	0,2538	0,0873	0,1130	0,1151	0,1535	0,1980	0,2833
0,39	0,1079	0,1239	0,1336	0,1511	0,2064	0,2638	0,0917	0,1183	0,1218	0,1604	0,2080	0,2969
0,395	0,1105	0,1270	0,1368	0,1544	0,2107	0,2689	0,0936	0,1210	0,1251	0,1639	0,2121	0,3036
0,397	0,1116	0,1281	0,1379	0,1558	0,2125	0,2708	0,0945	0,1121	0,1265	0,1653	0,2149	0,3062
0,4	0,1133	0,1299	0,1397	0,1581	0,2153	0,2738	0,0959	0,1238	0,1284	0,1675	0,2181	0,3102

$I^{(2)}(\tau, \theta, \psi)$

τ	0	15	30	45	60	75	0	15	30	45	60	75
0	0,4337	0,2728	0,1974	0,1757	0,1685	0,2475	0,2587	0,1573	0,1421	0,1284	0,1465	0,2184
0,1	0,3324	0,2111	0,1560	0,1416	0,1429	0,2254	0,2027	0,1314	0,1148	0,1080	0,1289	0,2087
0,2	0,2157	0,1433	0,1103	0,1033	0,1127	0,1905	0,1378	0,0912	0,0839	0,0941	0,1059	0,1865
0,25	0,1521	0,1066	0,0864	0,0830	0,0971	0,1680	0,1025	0,0729	0,0675	0,0713	0,0930	0,1710
0,3	0,0832	0,0680	0,0617	0,0619	0,0806	0,1416	0,0646	0,0544	0,0503	0,0578	0,0784	0,1507
0,32	0,0679	0,0550	0,0502	0,0503	0,0659	0,1180	0,0524	0,0441	0,0409	0,0471	0,0648	0,1261
0,34	0,0521	0,0423	0,0383	0,0386	0,0505	0,0921	0,0403	0,0339	0,0315	0,0361	0,0495	0,0986
0,36	0,0348	0,0278	0,0258	0,0264	0,0347	0,0641	0,0272	0,0228	0,0215	0,0246	0,0340	0,0686
0,38	0,0175	0,0142	0,0131	0,0134	0,0177	0,0334	0,0139	0,0116	0,0108	0,0125	0,0173	0,0359
0,39	0,0082	0,0068	0,0065	0,0068	0,0089	0,0170	0,0097	0,0057	0,0055	0,0062	0,0088	0,0182
0,395	0,0047	0,0036	0,0034	0,0034	0,0045	0,0084	0,0036	0,0030	0,0028	0,0031	0,0027	0,0092
0,397	0,0028	0,0024	0,0020	0,0018	0,0026	0,0050	0,0021	0,0017	0,0015	0,0020	0,0027	0,0057
0,0	0	0	0	0	0	0	0	0	0	0	0	0

Table I 77

TABLE I (continued)

60						75						c,°
0	15	30	45	60	75	0	15	30	45	60	75	θ,° / τ
0	0	0	0	0	0	0	0	0	0	0	0	0
0,0145	0,0147	0,0192	0,0252	0,0348	0,0631	0,0091	0,0094	0,0095	0,0124	0,0200	0,0395	0.1
0,0300	0,0303	0,0395	0,0512	0,0634	0,1139	0,0204	0,0210	0,0208	0,0267	0,0386	0,0763	0,2
0,0381	0,0384	0,0501	0,0645	0,0848	0,1358	0,0270	0,0276	0,0272	0,0347	0,0542	0,0965	0,25
0,0463	0,0465	0,0608	0,0778	0,1010	0,1558	0,0343	0,0350	0,0341	0,0433	0,0670	0,1131	0,3
0,0538	0,0548	0,0704	0,0899	0,1170	0,1801	0,0404	0,0415	0,0419	0,0529	0,0804	0,1354	0,32
0,0613	0,0631	0,0800	0,1022	0,1329	0,2034	0,0467	0,0484	0,0498	0,0629	0,0941	0,1578	0,34
0,0690	0,0714	0,0898	0,1143	0,1488	0,2262	0,0532	0,0555	0,0582	0,0732	0,1083	0,1803	0,36
0,0768	0,0799	0,0997	0,1267	0,1646	0,2481	0,0601	0,0629	0,0669	0,0840	0,1229	0,2031	0,38
0,0807	0,0841	0,1047	0,1329	0,1725	0,2588	0,0636	0,0668	0,0715	0,0896	0,1305	0,2147	0,39
0,0826	0,0863	0,1071	0,1359	0,1765	0,2641	0,0655	0,0688	0,0737	0,0924	0,1343	0,2207	0,395
0,0835	0,0871	0,1082	0,1371	0,1781	0,2664	0,0662	0,0695	0,0747	0,0936	0,1359	0,2230	0,397
0,0847	0,0885	0,1098	0,1389	0,1804	0,2695	0,0674	0,0707	0,0761	0,0953	0,1382	0,2266	0,4
0,1468	0,1228	0,1134	0,1141	,1518	0,2045	0,0848	0,0764	0,0778	0,0931	0,1140	0,1461	0
0,1201	0,1008	0,0944	0,0975	0,1337	0,1972	0,0741	0,0674	0,0692	0,0844	0,1078	0,1636	0,1
0,0879	0,0749	0,0720	0,0774	0,1085	0,1778	0,0590	0,0547	0,0572	0,0705	0,0948	0,1513	0,2
0,0703	0,0610	0,0600	0,0668	0,0938	0,1636	0,0496	0,0471	0,0498	0,0614	0,0855	0,1462	0,25
0,0514	0,0460	0,0473	0,0556	0,0773	0,1453	0,0388	0,0383	0,0413	0,0507	0,0738	0,1368	0,3
0,0420	0,0376	0,0386	0,0457	0,0638	0,1222	0,0322	0,0316	0,0343	0,0422	0,0618	0,1168	0,32
0,0322	0,0288	0,0297	0,0351	0,0493	0,0959	0,0249	0,0246	0,0266	0,0329	0,0485	0,0934	0,34
0,0220	0,0196	0,0203	0,0240	0,0338	0,0671	0,0172	0,0170	0,0185	0,0228	0,0339	0,0664	0,36
0,0111	0,0099	0,0103	0,0122	0,0173	0,0352	0,0090	0,0088	0,0096	0,0118	0,0177	0,0354	0,38
0,0057	0,0050	0,0051	0,0062	0,0088	0,0180	0,0045	0,0045	0,0049	0,0060	0,0091	0,0182	0,39
0,0028	0,0025	0,0026	0,0031	0,0044	0,0090	0,0023	0,0022	0,0025	0,0030	0,0045	0,0093	0,395
0,0018	0,0015	0,0016	0,0019	0,0026	0,0055	0,0014	0,0013	0,0015	0,0018	0,0027	0,0056	0,397
0	0	0	0	0	0	0	0	0	0	0	0	0,4
0	0	0	0	0	0	0	0	0	0	0	0	0
0,0145	0,0156	0,0201	0,0251	0,0413	0,0676	0,0091	0,0090	0,0099	0,0135	0,0213	0,0499	0,1
0,0300	0,0319	0,0414	0,0509	0,0765	0,1212	0,0204	0,0200	0,0218	0,0292	0,0407	0,0972	0,2
0,0381	0,0407	0,0524	0,0641	0,1016	0,1441	0,0270	0,0263	0,0284	0,0380	0,0573	0,1227	0,25
0,0463	0,0493	0,0637	0,0774	0,1212	0,1647	0,0343	0,0331	0,0358	0,0476	0,0707	0,1451	0,3
0,0538	0,0578	0,0754	0,0926	0,1394	0,1991	0,0404	0,0399	0,0441	0,0598	0,0888	0,1717	0,32
0,0613	0,0664	0,0870	0,1077	0,1575	0,2323	0,0467	0,0470	0,0525	0,0725	0,1072	0,1983	0,34
0,0690	0,0751	0,0990	0,1232	0,1754	0,2645	0,0532	0,0543	0,0615	0,0858	0,1265	0,2254	0,36
0,0768	0,0838	0,1108	0,1388	0,1935	0,2957	0,0601	0,0620	0,0709	0,0995	0,1465	0,2531	0,38
0,0807	0,0883	0,1167	0,1465	0,2026	0,3109	0,0636	0,0659	0,0757	0,1068	0,1567	0,2671	0,39
0,0826	0,0905	0,1197	0,1505	0,2070	0,3186	0,0655	0,0679	0,0782	0,1104	0,1620	0,2743	0,395
0,0835	0,0914	0,1210	0,1520	0,2088	0,3217	0,0662	0,0688	0,0792	0,1118	0,1641	0,2772	0,397
0,0847	0,0927	0,1230	0,1540	0,2115	0,3261	0,0674	0,0700	0,0807	0,1141	0,1673	0,2815	0,4
0,1468	0,1172	0,1010	0,1159	0,1568	0,2267	0,0848	0,0725	0,0797	0,0947	0,1245	0,1832	0
0,1201	0,0965	0,0847	0,0984	0,1375	0,2164	0,0741	0,0642	0,0709	0,0862	0,1175	0,2034	0,1
0,0879	0,0719	0,0658	0,0778	0,1120	0,1939	0,0590	0,0526	0,0583	0,0728	0,1036	0,1921	0,2
0,0703	0,0585	0,0561	0,0668	0,0980	0,1790	0,0496	0,0457	0,0504	0,0643	0,0939	0,1872	0,25
0,0514	0,0443	0,0460	0,0553	0,0828	0,1612	0,0388	0,0377	0,0414	0,0542	0,0821	0,1779	0,3
0,0420	0,0362	0,0377	0,0455	0,0683	0,1358	0,0322	0,0312	0,0343	0,0451	0,0688	0,1522	0,32
0,0322	0,0277	0,0290	0,0349	0,0529	0,1065	0,0249	0,0243	0,0266	0,0352	0,0541	0,1220	0,34
0,0220	0,0189	0,0197	0,0237	0,0362	0,0746	0,0172	0,0168	0,0185	0,0244	0,0377	0,0870	0,36
0,0111	0,0095	0,0100	0,0121	0,0186	0,0392	0,0090	0,0086	0,0096	0,0126	0,0197	0,0465	0,38
0,0057	0,0048	0,0051	0,0062	0,0094	0,0200	0,0045	0,0044	0,0049	0,0064	0,0100	0,0240	0,39
0,0028	0,0024	0,0026	0,0030	0,0047	0,0100	0,0023	0,0022	0,0025	0,0032	0,0050	0,0122	0,395
0,0018	0,0014	0,0015	0,0019	0,0028	0,0061	0,0014	0,0013	0,0015	0,0019	0,0031	0,0074	0,397
0	0	0	0	0	0	0	0	0	0	0	0	0,4

τ*=0,6;funct. VII; ψ=0° $\hfill I^{(1)}(\tau, \theta, \psi)$

ζ,°	30						45					
θ,° \ τ	0	15	30	45	60	75	0	15	30	45	60	75
0	0	0	0	0	0	0	0	0	0	0	0	0
0,1	0,0200	0,0174	0,0239	0,0319	0,0479	0,1311	0,0170	0,0150	0,0222	0,0298	0,0557	0,1528
0,2	0,0405	0,0349	0,0476	0,0631	0,0912	0,2290	0,0349	0,0307	0,0450	0,0595	0,1074	0,2698
0,3	0,0608	0,0524	0,0711	0,0926	0,1294	0,3008	0,0530	0,0464	0,0679	0,0882	0,1541	0,3586
0,4	0,0810	0,0692	0,0936	0,1201	0,1617	0,3502	0,0712	0,0618	0,0904	0,1151	0,1950	0,4230
0,45	0,0895	0,0759	0,1026	0,1301	0,1695	0,3489	0,0790	0,0678	0,0993	0,1249	0,2070	0,4268
0,50	0,1157	0,0982	0,1242	0,1529	0,1985	0,3767	0,0998	0,0879	0,1188	0,1476	0,2339	0,4545
0,55	0,1417	0,1201	0,1450	0,1747	0,2262	0,4011	0,1206	0,1079	0,1379	0,1698	0,2605	0,4808
0,57	0,1521	0,1290	0,1533	0,1832	0,2359	0,4105	0,1289	0,1158	0,1456	0,1786	0,2699	0,4908
0,58	0,1571	0,1334	0,1574	0,1874	0,2410	0,4147	0,1331	0,1198	0,1495	0,1831	0,2749	0,4960
0,59	0,1624	0,1373	0,1616	0,1917	0,2461	0,4190	0,1373	0,1238	0,1532	0,1875	0,2797	0,5010
0,595	0,1649	0,1397	0,1637	0,1936	0,2485	0,4212	0,1394	0,1260	0,1551	0,1896	0,2822	0,5037
0,6	0,1674	0,1420	0,1658	0,1957	0,2510	0,4232	0,1415	0,1279	0,1569	0,1919	0,2847	0,5060

$\hfill I^{(2)}(\tau, \theta, \psi)$

τ	0	15	30	45	60	75	0	15	30	45	60	75
0	0,5770	1,1393	1,6815	1,4389	0,9813	0,8382	0,3481	0,5579	1,1238	1,7400	1,6581	1,1675
0,1	0,5096	1,0017	1,4861	1,2875	0,9003	0,8161	0,3138	0,5003	1,0093	1,6298	1,5429	1,1544
0,2	0,4221	0,8272	1,2227	1,0713	0,7717	0,7500	0,2651	0,4202	0,8437	1,3771	1,3376	1,0728
0,3	0,3180	0,6054	0,8909	0,7986	0,6033	0,6407	0,2082	0,3116	0,6321	1,0319	1,0349	0,9108
0,4	0,1990	0,3386	0,4801	0,4558	0,3879	0,4759	0,1419	0,2031	0,3625	0,5758	0,6109	0,6376
0,45	0,1241	0,1778	0,2287	0,2375	0,2369	0,3228	0,0973	0,1267	0,1933	0,2826	0,3195	0,3971
0,50	0,0864	0,1249	0,1599	0,1669	0,1678	0,2383	0,0676	0,0887	0,1358	0,1920	0,2288	0,2958
0,55	0,0446	0,0640	0,0836	0,0880	0,0890	0,1313	0,0351	0,0460	0,0716	0,1017	0,1225	0,1646
0,57	0,0267	0,0387	0,0510	0,0538	0,0547	0,0820	0,0215	0,0280	0,0438	0,0624	0,0756	0,1031
0,58	0,0182	0,0260	0,0344	0,0364	0,0370	0,0558	0,0144	0,0189	0,0296	0,0422	0,0511	0,0704
0,59	0,0085	0,0125	0,0172	0,0185	0,0188	0,0285	0,0072	0,0094	0,0150	0,0213	0,0262	0,0360
0,595	0,0048	0,0067	0,0087	0,0092	0,0094	0,0142	0,0037	0,0048	0,0075	0,0108	0,0130	0,0180
0,6	0	0	0	0	0	0	0	0	0	0	0	0

τ*=0,6;funct. VII; ψ=45° $\hfill I^{(1)}(\tau, \theta, \psi)$

τ	0	15	30	45	60	75	0	15	30	45	60	75
0	0,0200	0,0188	0,0226	0,0314	0,0446	0,1139	0,0170	0,0153	0,0219	0,0276	0,0473	0,1216
0,1	0,0405	0,0379	0,0453	0,0621	0,0852	0,1992	0,0349	0,0312	0,0446	0,0551	0,0913	0,2149
0,3	0,0608	0,0567	0,0674	0,0914	0,1212	0,2617	0,0530	0,0471	0,0674	0,0818	0,1311	0,2855
0,4	0,0810	0,0751	0,0888	0,1189	0,1518	0,3051	0,0712	0,0629	0,0892	0,1070	0,1660	0,3364
0,45	0,0895	0,0826	0,0972	0,1290	0,1592	0,3027	0,0790	0,0691	0,0988	0,1160	0,1755	0,3365
0,5	0,1157	0,1056	0,1195	0,1522	0,1883	0,3350	0,0998	0,0896	0,1188	0,1387	0,2033	0,3660
0,55	0,1417	0,1280	0,1414	0,1746	0,2160	0,3631	0,1206	0,1097	0,1385	0,1609	0,2306	0,3926
0,57	0,1521	0,1370	0,1500	0,1835	0,2259	0,3736	0,1289	0,1180	0,1464	0,1696	0,2402	0,4028
0,58	0,1571	0,1415	0,1543	0,1877	0,2310	0,3785	0,1331	0,1221	0,1502	0,1741	0,2454	0,4078
0,59	0,1624	0,1459	0,1587	0,1921	0,2360	0,3834	0,1373	0,1261	0,1541	0,1783	0,2504	0,4126
0,595	0,1649	0,1482	0,1607	0,1943	0,2386	0,3858	0,1394	0,1282	0,1561	0,1805	0,2529	0,4152
0,6	0,1674	0,1504	0,1629	0,1964	0,2409	0,3881	0,1415	0,1302	0,1581	0,1827	0,2555	0,4174

$\hfill I^{(2)}(\tau, \theta, \psi)$

τ	0	15	30	45	60	75	0	15	30	45	60	75
0	0,5770	0,8476	0,9098	0,7578	0,6915	0,6192	0,3481	0,4636	0,5902	0,6757	0,7128	0,7122
0,1	0,5096	0,7466	0,8063	0,6795	0,6264	0,6054	0,3138	0,4164	0,5318	0,6191	0,6644	0,7054
0,2	0,4221	0,6180	0,6674	0,5688	0,5482	0,5605	0,2651	0,3505	0,4479	0,5276	0,5788	0,6587
0,3	0,3180	0,4585	0,4986	0,4357	0,4361	0,4900	0,2082	0,2707	0,3453	0,4117	0,4631	0,5734
0,4	0,1990	0,2693	0,2964	0,2766	0,2970	0,3888	0,1419	0,1762	0,2206	0,2689	0,3137	0,4390
0,45	0,1241	0,1540	0,1697	0,1712	0,1049	0,2840	0,0973	0,1142	0,1391	0,1716	0,2026	0,3057
0,5	0,0864	0,1080	0,1184	0,1198	0,1378	0,2092	0,0676	0,0798	0,0974	0,1163	0,1441	0,2267
0,55	0,0446	0,0552	0,0617	0,0628	0,0729	0,1151	0,0354	0,0414	0,0511	0,0612	0,0767	0,1257
0,57	0,0267	0,0334	0,0376	0,0383	0,0447	0,0717	0,0213	0,0252	0,0311	0,0375	0,0471	0 0786
0,58	0,0182	0,0224	0,0253	0,0260	0,0302	0,0489	0,0144	0,0170	0,0211	0,0253	0,0318	0,0535
0,59	0,0085	0,0108	0,0127	0,0133	0,0154	0,0250	0,0072	0,0084	0,0107	0,0127	0,0163	0,0274
0,595	0,0048	0,0058	0,0063	0,0066	0,0077	0,0124	0,0037	0,0044	0,0053	0,0064	0,0081	0,0137
0,6	0	0	0	0	0	0	0	0	0	0	0	0

Table I 79

TABLE I (continued)

60						75						ζ,°
0	15	30	45	60	75	0	15	30	45	60	75	θ,° / τ
0	0	0	0	0	0	0	0	0	0	0	0	0
0,0131	0,0146	0,0172	0,0286	0,0585	0,1780	0,0066	0,0073	0,0108	0,0176	0,0386	0,1127	0,1
0,0274	0,0306	0,0356	0,0586	0,1160	0,3257	0,0146	0,0162	0,0239	0,0386	0,0834	0,2272	0,2
0,0426	0,0474	0,0546	0,0893	0,1718	0,4511	0,0243	0,0269	0,0400	0,0641	0,1312	0,3530	0,3
0,0585	0,0652	0,0740	0,1197	0,2248	0,5579	0,0361	0,0397	0,0595	0,0949	0,2011	0,5003	0,4
0,0655	0,0730	0,0819	0,1318	0,2436	0,5836	0,0419	0,0462	0,0695	0,1106	0,2337	0,5684	0,45
0,0837	0,0902	0,1012	0,1540	0,2724	0,5969	0,0557	0,0607	0,0858	0,1323	0,2611	0,5831	0,50
0,1022	0,1076	0,1207	0,1764	0,3019	0,6149	0,0711	0,0767	0,1040	0,1569	0,2938	0,6133	0,55
0,1098	0,1148	0,1286	0,1855	0,3132	0,6234	0,0777	0,0837	0,1118	0,1677	0,3078	0,6299	0,57
0,1137	0,1184	0,1326	0,1900	0,3190	0,6280	0,0811	0,0873	0,1159	0,1732	0,3153	0,6392	0,58
0,1175	0,1220	0,1366	0,1946	0,3250	0,6327	0,0837	0,0910	0,1201	0,1789	0,3232	0,6491	0,59
0,1200	0,1239	0,1386	0,1968	0,3279	0,6353	0,0865	0,0929	0,1223	0,1818	0,3273	0,6544	0,595
0,1214	0,1256	0,1407	0,1991	0,3311	0,6378	0,0883	0,0948	0,1244	0,1848	0,3314	0,6599	0,6
0,1976	0,3034	0,5056	1,0864	1,8194	1,7414	0,1057	0,1338	0,2062	0,3463	0,7503	1,2025	0
0,1846	0,2814	0,4689	1,0162	1,7469	1,7958	0,1043	0,1323	0,2046	0,3484	0,7804	1,3746	0,1
0,1619	0,2445	0,4061	0,8855	1,5586	1,7288	0,0936	0,1227	0,1902	0,3276	0,7569	1,4627	0,2
0,1344	0,1974	0,3219	0,6901	1,2320	1,4879	0,0856	0,1082	0,1655	0,2839	0,6610	1,3836	0,3
0,1019	0,1384	0,2112	0,4137	0,7213	0,9733	0,0700	0,0872	0,1263	0,2069	0,4503	0,9706	0,4
0,0769	0,0964	0,1348	0,2243	0,3542	0,5200	0,0554	0,0679	0,0922	0,1404	0,2640	0,5178	0,45
0,0540	0,0679	0,0955	0,1606	0,2577	0,3945	0,0401	0,0495	0,0677	0,1043	0,2600	0,4120	0,50
0,0282	0,0357	0,0507	0,0860	0,1403	0,2241	0,0218	0,0270	0,0372	0,0580	0,1135	0,2457	0,55
0,0173	0,0218	0,0311	0,0530	0,0871	0,1414	0,0136	0,0167	0,0232	0,0363	0,0716	0,1581	0,57
0,0116	0,0147	0,0209	0,0358	0,0591	0,0970	0,0092	0,0114	0,0158	0,0247	0,0491	0,1093	0,58
0,0059	0,0074	0,0107	0,0184	0,0301	0,0496	0,0047	0,0058	0,0081	0,0126	0,0251	0,0566	0,59
0,0029	0,0038	0,0053	0,0092	0,0152	0,0250	0,0024	0,0029	0,0041	0,0063	0,0127	0,0288	0,595
0	0	0	0	0	0	0	0	0	0	0	0	0,6
0	0	0	0	0	0	0	0	0	0	0	0	0
0,0131	0,0146	0,0161	0,0243	0,0460	0,1113	0,0066	0,0070	0,0096	0,0144	0,0253	0,0650	0,1
0,0274	0,0307	0,0334	0,0498	0,0913	0,2020	0,0146	0,0155	0,0213	0,0315	0,0538	0,1275	0,2
0,0426	0,0477	0,0513	0,0756	0,1350	0,2760	0,0243	0,0257	0,0353	0,0519	0,0822	0,1910	0,3
0,0585	0,0655	0,0696	0,1012	0,1762	0,3345	0,0361	0,0378	0,0523	0,0761	0,1228	0,2585	0,4
0,0655	0,0735	0,0771	0,1112	0,1898	0,3393	0,0419	0,0439	0,0607	0,0879	0,1387	0,2798	0,45
0,0837	0,0908	0,0962	0,1322	0,2161	0,3727	0,0557	0,0583	0,0766	0,1070	0,1653	0,3139	0,5
0,1022	0,1084	0,1155	0,1555	0,2427	0,4053	0,0711	0,0742	0,0943	0,1283	0,1953	0,3537	0,55
0,1098	0,1157	0,1232	0,1644	0,2526	0,4183	0,0777	0,0811	0,1019	0,1376	0,2077	0,3716	0,57
0,1137	0,1193	0,1271	0,1689	0,2578	0,4248	0,0810	0,0847	0,1059	0,1424	0,2114	0,3811	0,58
0,1175	0,1230	0,1312	0,1736	0,2631	0,4313	0,0847	0,0883	0,1099	0,1472	0,2212	0,3907	0,59
0,1200	0,1249	0,1331	0,1757	0,2657	0,4346	0,0865	0,0902	0,1121	0,1497	0,2248	0,3958	0,595
0,1214	0,1266	0,1351	0,1780	0,2684	0,4379	0,0883	0,0920	0,1141	0,1523	0,2283	0,4008	0,6
0,1976	0,2659	0,3597	0,4702	0,5828	0,6702	0,1057	0,1224	0,1610	0,2242	0,3081	0,3781	0
0,1846	0,2472	0,3347	0,4414	0,5604	0,6869	0,1043	0,1210	0,1601	0,2257	0,3193	0,4243	0,1
0,1619	0,2158	0,2913	0,3869	0,5024	0,6605	0,0966	0,1122	0,1492	0,2124	0,3087	0,4457	0,2
0,1344	0,1758	0,2246	0,3122	0,4144	0,5879	0,0856	0,0930	0,1318	0,1872	0,2775	0,4344	0,3
0,1019	0,1266	0,1625	0,2140	0,2907	0,4510	0,0700	0,0807	0,1053	0,1459	0,2165	0,3660	0,4
0,0769	0,0908	0,1107	0,1418	0,1921	0,3079	0,0554	0,0634	0,0807	0,1073	0,1547	0,2644	0,45
0,0540	0,0639	0,0782	0,1009	0,1385	0,2317	0,0401	0,0461	0,0592	0,0793	0,1161	0,2077	0,5
0,0282	0,0336	0,0413	0,0536	0,0747	0,1303	0,0218	0,0251	0,0324	0,0439	0,0651	0,1221	0,55
0,0173	0,0205	0,0253	0,0329	0,0462	0,0819	0,0136	0,0156	0,0202	0,0274	0,0409	0,0782	0,57
0,0116	0,0138	0,0170	0,0222	0,0313	0,0560	0,0092	0,0106	0,0138	0,0186	0,0280	0,0539	0,58
0,0059	0,0070	0,0087	0,0114	0,0159	0,0287	0,0047	0,0054	0,0070	0,0095	0,0143	0,0278	0,59
0,0029	0,0036	0,0043	0,0057	0,0080	0,0144	0,0024	0,0027	0,0035	0,0047	0,0072	0,0141	0,595
0	0	0	0	0	0	0	0	0	0	0	0	0,6

τ*=0,6; funct.VII; ψ=90° $I^{(1)}(\tau, \theta, \psi)$

ζ,°	30						45					
θ,° τ	0	15	30	45	60	75	0	15	30	45	60	75
0	0	0	0	0	0	0	0	0	0	0	0	0
0,1	0,0200	0,0213	0,0236	0,0274	0,0441	0,0895	0,0170	0,0175	0,0193	0,0257	0,0409	0,0811
0,2	0,0405	0,0429	0,0471	0,0543	0,0846	0,1572	0,0349	0,0357	0,0392	0,0513	0,0795	0,1440
0,3	0,0608	0,0644	0,0704	0,0799	0,1211	0,2076	0,0530	0,0542	0,0593	0,0766	0,1152	0,1923
0,4	0,0810	0,0857	0,0929	0,1039	0,1533	0,2433	0,0712	0,0726	0,0788	0,1007	0,1475	0,2281
0,45	0,0895	0,0946	0,1020	0,1124	0,1622	0,2402	0,0790	0,0802	0,0866	0,1095	0,1569	0,2257
0,5	0,1157	0,1202	0,1267	0,1392	0,1927	0,2799	0,0998	0,1014	0,1091	0,1345	0,1859	0,2659
0,55	0,1417	0,1453	0,1510	0,1651	0,2220	0,3140	0,1206	0,1225	0,1315	0,1590	0,2139	0,3012
0,57	0,1521	0,1555	0,1605	0,1751	0,2324	0,3165	0,1289	0,1310	0,1402	0,1685	0,2239	0,3142
0,58	0,1571	0,1605	0,1653	0,1802	0,2378	0,3323	0,1331	0,1352	0,1447	0,1734	0,2292	0,3206
0,59	0,1624	0,1654	0,1702	0,1852	0,2431	0,3380	0,1373	0,1394	0,1491	0,1781	0,2345	0,3268
0,595	0,1649	0,1680	0,1725	0,1877	0,2457	0,3409	0,1394	0,1416	0,1514	0,1806	0,2370	0,3298
0,6	0,1674	0,1705	0,1748	0,1902	0,2483	0,3438	0,1415	0,1437	0,1535	0,1829	0,2398	0,3326

$I^{(2)}(\tau, \theta, \psi)$

τ	0	15	30	45	60	75	0	15	30	45	60	75
0	0,5770	0,5326	0,4674	0,4097	0,3769	0,4107	0,3481	0,3414	0,3233	0,3039	0,3224	0,3657
0,1	0,5096	0,4707	0,4168	0,3717	0,3522	0,4067	0,3138	0,3084	0,2944	0,2853	0,3068	0,3697
0,2	0,4221	0,3919	0,3486	0,3162	0,3095	0,3826	0,2651	0,2615	0,2517	0,2488	0,2743	0,3543
0,3	0,3180	0,2974	0,2694	0,2524	0,2582	0,3447	0,2082	0,2061	0,2016	0,2054	0,2328	0,3252
0,4	0,1990	0,1890	0,1786	0,1791	0,1982	0,2911	0,1419	0,1419	0,1436	0,1553	0,1823	0,2802
0,45	0,1241	0,1209	0,1186	0,1263	0,1469	0,2230	0,0973	0,0983	0,1025	0,1159	0,1362	0,2170
0,5	0,0864	0,0846	0,0824	0,0879	0,1035	0,1638	0,0676	0,0685	0,0714	0,0783	0,0964	0,1602
0,55	0,0446	0,0431	0,0427	0,0459	0,0545	0,0896	0,0351	0,0355	0,0373	0,0409	0,0508	0,0882
0,57	0,0267	0,0261	0,0260	0,0280	0,0334	0,0558	0,0213	0,0215	0,0227	0,0250	0,0311	0,0549
0,58	0,0182	0,0175	0,0174	0,0190	0,0225	0,0380	0,0144	0,0145	0,0153	0,0168	0,0210	0,0373
0,59	0,0085	0,0084	0,0088	0,0097	0,0115	0,0193	0,0072	0,0072	0,0078	0,0085	0,0107	0,0191
0,595	0,0048	0,0045	0,0043	0,0048	0,0057	0,0097	0,0037	0,0037	0,0039	0,0043	0,0053	0,0095
0,6	0	0	0	0	0	0	0	0	0	0	0	0

τ*=0,6; funct.VII; ψ=135° $I^{(1)}(\tau, \theta, \psi)$

τ	0	15	30	45	60	75	0	15	30	45	60	75
0	0	0	0	0	0	0	0	0	0	0	0	0
0,1	0,0200	0,0217	0,0217	0,0296	0,0454	0,0787	0,0170	0,0209	0,0209	0,0273	0,0421	0,0721
0,2	0,0405	0,0436	0,0435	0,0586	0,0875	0,1386	0,0349	0,0426	0,0425	0,0548	0,0823	0,1288
0,3	0,0608	0,0655	0,0649	0,0865	0,1257	0,1835	0,0530	0,0649	0,0644	0,0818	0,1199	0,1729
0,4	0,0810	0,0871	0,0855	0,1129	0,1599	0,2156	0,0712	0,0874	0,0860	0,1081	0,1547	0,2064
0,45	0,0895	0,0963	0,0937	0,1226	0,1701	0,2120	0,0790	0,0973	0,0947	0,1182	0,1656	0,2043
0,5	0,1157	0,1254	0,1264	0,1574	0,2059	0,2635	0,0998	0,1210	0,1237	0,1516	0,2032	0,2581
0,55	0,1417	0,1544	0,1585	0,1910	0,2403	0,3078	0,1206	0,1446	0,1526	0,1844	0,2398	0,3053
0,57	0,1521	0,1660	0,1712	0,2043	0,2528	0,3238	0,1289	0,1543	0,1640	0,1974	0,2533	0,3228
0,58	0,1571	0,1717	0,1775	0,2109	0,2592	0,3316	0,1331	0,1591	0,1697	0,2039	0,2601	0,3312
0,59	0,1624	0,1774	0,1839	0,2175	0,2656	0,3391	0,1373	0,1638	0,1755	0,2104	0,2671	0,3395
0,595	0,1649	0,1803	0,1871	0,2207	0,2687	0,3428	0,1394	0,1664	0,1784	0,2138	0,2705	0,3436
0,6	0,1674	0,1832	0,1903	0,2239	0,2719	0,3465	0,1415	0,1687	0,1813	0,2169	0,2741	0,3476

$I^{(2)}(\tau, \theta, \psi)$

τ	0	15	30	45	60	75	0	15	30	45	60	75
0	0,5770	0,4138	0,3262	0,2813	0,2864	0,3247	0,3481	0,2533	0,2236	0,2238	0,2293	0,3001
0,1	0,5096	0,3669	0,2932	0,2574	0,2707	0,3261	0,3138	0,2303	0,2054	0,2096	0,2223	0,3083
0,2	0,4221	0,3069	0,2478	0,2213	0,2408	0,3121	0,2651	0,1972	0,1772	0,1847	0,2031	0,3011
0,3	0,3180	0,2364	0,1968	0,1810	0,2056	0,2903	0,2082	0,1594	0,1453	0,1555	0,1792	0,2845
0,4	0,1990	0,1572	0,1402	0,1365	0,1652	0,2601	0,1419	0,1173	0,1098	0,1227	0,1516	0,2572
0,45	0,1241	0,1060	0,1006	0,1014	0,1269	0,2100	0,0973	0,0871	0,0823	0,0943	0,1208	0,2078
0,5	0,0864	0,0741	0,0698	0,0704	0,0892	0,1542	0,0676	0,0606	0,0571	0,0637	0,0853	0,1534
0,55	0,0446	0,0377	0,0361	0,0366	0,0467	0,0843	0,0351	0,0313	0,0296	0,0331	0,0449	0,0844
0,57	0,0267	0,0227	0,0219	0,0222	0,0286	0,0524	0,0213	0,0190	0,0180	0,0202	0,0274	0,0526
0,58	0,0182	0,0153	0,0147	0,0150	0,0193	0,0357	0,0144	0,0128	0,0121	0,0136	0,0185	0,0359
0,59	0,0085	0,0074	0,0074	0,0077	0,0099	0,0181	0,0072	0,0064	0,0061	0,0068	0,0094	0,0183
0,595	0,0048	0,0040	0,0036	0,0038	0,0049	0,0091	0,0037	0,0033	0,0030	0,0034	0,0047	0,0092
0,6	0	0	0	0	0	0	0	0	0	0	0	0

Table I 81

TABLE I (continued)

60						75						ζ,°
0	15	30	45	60	75	0	15	30	45	60	75	0,° / τ
0	0	0	0	0	0	0	0	0	0	0	0	0
0,0131	0,0138	0,0163	0,0214	0,0323	0,0646	0,0066	0,0069	0,0080	0,0105	0,0162	0,0341	0,1
0,0274	0,0289	0,0339	0,0440	0,0642	0,1178	0,0146	0,0154	0,0176	0,0226	0,0340	0,0655	0,2
0,0426	0,0449	0,0523	0,0673	0,0952	0,1615	0,0243	0,0255	0,0290	0,0367	0,0534	0,0955	0,3
0,0585	0,0616	0,0715	0,0907	0,1249	0,1967	0,0361	0,0377	0,0425	0,0528	0,0747	0,1250	0,4
0,0655	0,0690	0,0796	0,1000	0,1340	0,1968	0,0419	0,0438	0,0489	0,0600	0,0827	0,1293	0,45
0,0837	0,0875	0,0995	0,1222	0,1616	0,2373	0,0557	0,0580	0,0645	0,0785	0,1067	0,1665	0,5
0,1022	0,1063	0,1195	0,1447	0,1892	0,2743	0,0711	0,0737	0,0818	0,0987	0,1331	0,2048	0,55
0,1099	0,1140	0,1278	0,1537	0,1993	0,2884	0,0777	0,0805	0,0892	0,1074	0,1437	0,2207	0,57
0,1137	0,1179	0,1317	0,1583	0,2046	0,2954	0,0811	0,0841	0,0930	0,1118	0,1493	0,2288	0,58
0,1175	0,1219	0,1360	0,1628	0,2100	0,3023	0,0847	0,0877	0,0970	0,1164	0,1551	0,2370	0,59
0,1200	0,1238	0,1381	0,1651	0,2127	0,3057	0,0865	0,0896	0,0991	0,1187	0,1581	0,2411	0,595
0,1214	0,1257	0,1402	0,1673	0,2154	0,3090	0,0883	0,0914	0,1011	0,1211	0,1610	0,2454	0,6
0,1976	0,1986	0,2041	0,2211	0,2545	0,2887	0,1057	0,1079	0,1163	0,1313	0,1528	0,1688	0
0,1846	0,1857	0,1922	0,2111	0,2501	0,3032	0,1043	0,1067	0,1159	0,1330	0,1598	0,1907	0,1
0,1619	0,1633	0,1698	0,1888	0,2302	0,3008	0,0966	0,0989	0,1082	0,1258	0,1562	0,2035	0,2
0,1344	0,1359	0,1424	0,1603	0,2012	0,2867	0,0856	0,0876	0,0965	0,1137	0,1462	0,2095	0,3
0,1019	0,1034	0,1096	0,1254	0,1625	0,2571	0,0700	0,0718	0,0797	0,0953	0,1273	0,2040	0,4
0,0769	0,0782	0,0832	0,0949	0,1234	0,2046	0,0554	0,0566	0,0628	0,0752	0,1016	0,1709	0,45
0,0540	0,0550	0,0585	0,0671	0,0884	0,1528	0,0401	0,0410	0,0457	0,0551	0,0755	0,1329	0,5
0,0282	0,0288	0,0307	0,0353	0,0472	0,0853	0,0218	0,0223	0,0249	0,0302	0,0419	0,0773	0,55
0,0173	0,0175	0,0188	0,0216	0,0291	0,0534	0,0136	0,0138	0,0155	0,0188	0,0262	0,0493	0,57
0,0116	0,0118	0,0126	0,0146	0,0196	0,0364	0,0092	0,0093	0,0105	0,0127	0,0179	0,0339	0,58
0,0059	0,0060	0,0064	0,0074	0,0100	0,0186	0,0047	0,0047	0,0053	0,0064	0,0091	0,0175	0,59
0,0029	0,0031	0,0032	0,0037	0,0050	0,0094	0,0024	0,0024	0,0027	0,0032	0,0046	0,0088	0,595
0	0	0	0	0	0	0	0	0	0	0	0	0,6
0	0	0	0	0	0	0	0	0	0	0	0	0
0,0131	0,0133	0,0170	0,0224	0,0320	0,0611	0,0066	0,0068	0,0073	0,0097	0,0156	0,0322	0,1
0,0274	0,0280	0,0353	0,0460	0,0639	0,1122	0,0146	0,0152	0,0160	0,0207	0,0327	0,0621	0,2
0,0426	0,0434	0,0547	0,0704	0,0953	0,1551	0,0243	0,0251	0,0261	0,0334	0,0553	0,0910	0,3
0,0585	0,0594	0,0750	0,0956	0,1258	0,1912	0,0361	0,0370	0,0377	0,0479	0,0726	0,1204	0,4
0,0655	0,0665	0,0839	0,1059	0,1358	0,1931	0,0419	0,0430	0,0432	0,0542	0,0807	0,1257	0,45
0,0837	0,0862	0,1067	0,1341	0,1722	0,2461	0,0557	0,0579	0,0696	0,0758	0,1099	0,1729	0,5
0,1022	0,1063	0,1299	0,1626	0,2090	0,2948	0,0711	0,0743	0,0799	0,0995	0,1420	0,2214	0,55
0,1098	0,1145	,1394	0,1742	0,2227	0,3134	0,0777	0,0815	0,0882	0,1097	0,1551	0,2414	0,57
0,1137	0,1187	,1441	0,1800	0,2299	0,3225	0,0811	0,0852	0,0925	0,1150	0,1620	0,2517	0,58
0,1175	0,1229	0,1490	0,1859	0,2371	0,3315	0,0847	0,0891	0,0969	0,1203	0,1691	0,2620	0,59
0,1200	0,1250	0,1514	0,1888	0,2408	0,3360	0,0865	0,0910	0,0992	0,1230	0,1728	0,2673	0,595
0,1214	0,1271	0,1538	0,1917	0,2444	0,3404	0,0883	0,0929	0,1015	0,1259	0,1764	0,2726	0,6
0,1976	0,1691	0,1591	0,1615	0,2016	0,2391	0,1057	0,0972	0,0997	0,1155	0,1350	0,1495	0
0,1846	0,1587	0,1511	0,1565	0,2013	0,2557	0,1043	0,0964	0,0999	0,1177	0,1427	0,1713	0,1
0,1619	0,1398	0,1340	0,1425	0,1887	0,2598	0,0966	0,0897	0,0938	0,1123	0,1416	0,1865	0,2
0,1344	0,1171	0,1147	0,1251	0,1699	0,2570	0,0856	0,0803	0,0849	0,1031	0,1358	0,1986	0,3
0,1019	0,0904	0,0918	0,1048	0,1448	0,2454	0,0700	0,0672	0,0720	0,0889	0,1237	0,2052	0,4
0,0769	0,0691	0,0717	0,0842	0,1155	0,2073	0,0554	0,0542	0,0590	0,0721	0,1032	0,1826	0,45
0,0540	0,0485	0,0502	0,0594	0,0827	0,1551	0,0401	0,0393	0,0429	0,0528	0,0768	0,1424	0,5
0,0282	0,0253	0,0263	0,0312	0,0441	0,0866	0,0218	0,0213	0,0233	0,0289	0,0427	0,0831	0,55
0,0173	0,0153	0,0161	0,0190	0,0271	0,0543	0,0136	0,0132	0,0145	0,0180	0,0267	0,0530	0,57
0,0116	0,0104	0,0108	0,0128	0,0183	0,0371	0,0092	0,0089	0,0099	0,0122	0,0182	0,0365	0,58
0,0059	0,0053	0,0055	0,0066	0,0093	0,0189	0,0047	0,0045	0,0050	0,0061	0,0093	0,0188	0,59
0,0029	0,0027	0,0027	0,0033	0,0047	0,0096	0,0024	0,0023	0,0025	0,0031	0,0047	0,0095	0,595
0	0	0	0	0	0	0	0	0	0	0	0	0,6

$\tau^{*}=0,6$; funct. VII; $\psi=180°$ $I^{(1)}(\tau,\ \theta,\ \psi)$

$\zeta,°$	30						45					
$\theta,°$ $\backslash$ τ	0	15	30	45	60	75	0	15	30	45	60	75
0	0	0	0	0	0	0	0	0	0	0	0	0
0,1	0,0200	0,0231	0,0265	0,0286	0,0468	0,0807	0,0170	0,0212	0,0204	0,0315	0,0425	0,0777
0,2	0,0405	0,0466	0,0529	0,0569	0,0901	0,1423	0,0349	0,0434	0,0416	0,0631	0,0828	0,1388
0,3	0,0608	0,0700	0,0794	0,0839	0,1296	0,1885	0,0530	0,0662	0,0628	0,0945	0,1207	0,1867
0,4	0,0810	0,0931	0,1050	0,1092	0,1651	0,2221	0,0712	0,0892	0,0833	0,1255	0,1554	0,2232
0,45	0,0895	0,1032	0,1158	0,1185	0,1759	0,2192	0,0790	0,0994	0,0923	0,1379	0,1662	0,2218
0,5	0,1157	0,1322	0,1451	0,1572	0,2202	0,2748	0,0998	0,1253	0,1242	0,1712	0,2142	0,2931
0,55	0,1417	0,1607	0,1733	0,1945	0,2626	0,3227	0,1206	0,1511	0,1559	0,2040	0,2611	0,3562
0,57	0,1521	0,1722	0,1850	0,2093	0,2783	0,3403	0,1289	0,1617	0,1687	0,2170	0,2786	0,3796
0,58	0,1571	0,1781	0,1908	0,2165	0,2862	0,3486	0,1331	0,1669	0,1749	0,2234	0,2876	0,3908
0,59	0,1624	0,1836	0,1965	0,2240	0,2942	0,3563	0,1373	0,1721	0,1813	0,2370	0,2965	0,4019
0,595	0,1649	0,1865	0,1994	0,2274	0,2982	0,3608	0,1394	0,1748	0,1845	0,2333	0,3010	0,4075
0,6	0,1674	0,1894	0,2021	0,2310	0,3020	0,3648	0,1415	0,1775	0,1878	0,2365	0,3055	0,4129

$I^{(2)}(\tau,\ \theta,\ \psi)$

τ	0	15	30	45	60	75	0	15	30	45	60	75
0	0,5770	0,3788	0,2896	0,2674	0,2606	0,3317	0,3481	0,2262	0,2114	0,2129	0,2356	0,3049
0,1	0,5096	0,3363	0,2607	0,2442	0,2468	0,3329	0,3138	0,2063	0,1942	0,1938	0,2272	0,3129
0,2	0,4221	0,2819	0,2206	0,2097	0,2205	0,3185	0,2651	0,1772	0,1675	0,1709	0,2063	0,3057
0,3	0,3180	0,2186	0,1765	0,1712	0,1906	0,2950	0,2082	0,1447	0,1373	0,1450	0,1809	0,2903
0,4	0,1990	0,1481	0,1284	0,1292	0,1577	0,2621	0,1419	0,1090	0,1038	0,1171	0,1517	0,2671
0,45	0,1241	0,1020	0,0939	0,0959	0,1240	0,2100	0,0973	0,0826	0,0778	0,0919	0,1202	0,2204
0,5	0,0864	0,0711	0,0651	0,0666	0,0871	0,1541	0,0676	0,0575	0,0540	0,0622	0,0849	0,1630
0,55	0,0446	0,0362	0,0333	0,0345	0,0458	0,0844	0,0351	0,0297	0,0280	0,0321	0,0446	0,0898
0,57	0,0267	0,0218	0,0304	0,0210	0,0280	0,0524	0,0213	0,0180	0,0170	0,0197	0,0274	0,0560
0,58	0,0182	0,0147	0,0136	0,0141	0,0190	0,0357	0,0144	0,0122	0,0113	0,0133	0,0185	0,0382
0,59	0,0085	0,0071	0,0069	0,0072	0,0096	0,0182	0,0072	0,0060	0,0058	0,0066	0,0093	0,0195
0,595	0,0048	0,0037	0,0034	0,0036	0,0048	0,0091	0,0037	0,0031	0,0028	0,0033	0,0046	0,0097
0,6	0	0	0	0	0	0	0	0	0	0	0	0

$\tau^{*}=0,6$; funct. VIII; $\psi=0°$ $I^{(1)}(\tau,\ \theta,\ \psi)$

τ	0	15	30	45	60	75	0	15	30	45	60	75
0	0	0	0	0	0	0	0	0	0	0	0	0
0,1	0,0121	0,0086	0,0162	0,0236	0,0345	0,1196	0,0104	0,0074	0,0160	0,0215	0,0463	0,1457
0,2	0,0245	0,0170	0,0325	0,0466	0,0654	0,2072	0,0211	0,0152	0,0323	0,0429	0,0887	0,2541
0,3	0,0368	0,0254	0,0483	0,0684	0,0919	0,2693	0,0321	0,0225	0,0486	0,0632	0,1260	0,3320
0,4	0,0489	0,0331	0,0636	0,0884	0,1133	0,3095	0,0431	0,0293	0,0645	0,0815	0,1571	0,3827
0,45	0,0541	0,0357	0,0698	0,0958	0,1171	0,3058	0,0477	0,0317	0,0710	0,0878	0,1652	0,3800
0,5	0,0807	0,0587	0,0917	0,1193	0,1488	0,3368	0,0689	0,0524	0,0906	0,1115	0,1938	0,4117
0,55	0,1072	0,0816	0,1130	0,1416	0,1788	0,3641	0,0901	0,0730	0,1102	0,1346	0,2219	0,4415
0,57	0,1177	0,0906	0,1215	0,1504	0,1897	0,3745	0,0986	0,0812	0,1178	0,1438	0,2321	0,4532
0,58	0,1230	0,0951	0,1257	0,1548	0,1953	0,3793	0,1029	0,0853	0,1220	0,1484	0,2374	0,4587
0,59	0,1282	0,0995	0,1301	0,1590	0,2008	0,3841	0,1072	0,0896	0,1258	0,1530	0,2428	0,4645
0,595	0,1309	0,1018	0,1322	0,1613	0,2033	0,3865	0,1093	0,0917	0,1278	0,1554	0,2453	0,4673
0,6	0,1335	0,1041	0,1342	0,1634	0,2063	0,3888	0,1115	0,0937	0,1295	0,1576	0,2480	0,4702

$I^{(2)}(\tau,\ \theta,\ \psi)$

τ	0	15	30	45	60	75	0	15	30	45	60	75
0	0,6542	1,4402	2,2138	1,8333	1,1492	0,9141	0,3581	0,6319	1,4138	2,3576	2,1230	1,3854
0,1	0,5704	1,2537	1,9354	1,6216	1,0377	0,8727	0,3191	0,5601	1,2571	2,1200	1,9531	1,3456
0,2	0,4655	1,0219	1,5714	1,3300	0,8727	0,7857	0,2669	0,4641	1,0375	1,7688	1,6927	1,2248
0,3	0,3432	0,7298	1,1195	0,9675	0,6644	0,6559	0,2071	0,3481	0,7591	1,2942	1,2805	1,0108
0,4	0,2056	0,3812	0,5615	0,5146	0,4065	0,4730	0,1391	0,2105	0,4068	0,6722	0,7077	0,6705
0,45	0,1212	0,1748	0,2252	0,2331	0,2308	0,3124	0,0946	0,1240	0,1900	0,2788	0,3140	0,3873
0,5	0,0846	0,1230	0,1577	0,1640	0,1640	0,2311	0,0659	0,0870	0,1337	0,1895	0,2252	0,2890
0,55	0,0437	0,0632	0,0826	0,0866	0,0870	0,1296	0,0342	0,0452	0,0706	0,1005	0,1208	0,1613
0,57	0,0262	0,0381	0,0503	0,0530	0,0536	0,0798	0,0207	0,0275	0,0431	0,0617	0,0746	0,1011
0,58	0,0178	0,0257	0,0340	0,0359	0,0362	0,0543	0,0141	0,0187	0,0292	0,0417	0,0503	0,0690
0,59	0,0083	0,0122	0,0170	0,0182	0,0184	0,0278	0,0070	0,0093	0,0149	0,0210	0,0258	0,0352
0,595	0,0047	0,0067	0,0085	0,0091	0,0092	0,0138	0,0037	0,0048	0,0074	0,0107	0,0129	0,0177
0,6	0	0	0	0	0	0	0	0	0	0	0	0

Table I 83

TABLE I(continued)

60						75						ζ,°
0	15	30	45	60	75	0	15	30	45	60	75	θ,° / τ
0	0	0	0	0	0	0	0	0	0	0	0	0
0,0131	0,0140	0,0176	0,0223	0,0366	0,0651	0,0066	0,0067	0,0075	0,0102	0,0164	0,0380	0,1
0,0274	0,0292	0,0366	0,0459	0,0733	0,1192	0,0146	0,0147	0,0164	0,0219	0,0344	0,0738	0,2
0,0426	0,0454	0,0567	0,0704	0,1097	0,1643	0,0243	0,0243	0,0268	0,0356	0,0595	0,1093	0,3
0,0585	0,0623	0,0779	0,0955	0,1455	0,2015	0,0361	0,0356	0,0390	0,0513	0,0761	0,1462	0,4
0,0655	0,0698	0,0872	0,1058	0,1581	0,2027	0,0419	0,0412	0,0448	0,0584	0,0845	0,1549	0,45
0,0837	0,0902	0,1146	0,1410	0,1992	0,2781	0,0557	0,0565	0,0633	0,0855	0,1237	0,2108	0,5
0,1022	0,1109	0,1426	0,1767	0,2406	0,3475	0,0711	0,0735	0,0839	0,1155	0,1668	0,2690	0,55
0,1098	0,1195	0,1540	0,1914	0,2562	0,3742	0,0777	0,0809	0,0928	0,1286	0,1848	0,2933	0,57
0,1137	0,1238	0,1597	0,1986	0,2645	0,3872	0,0811	0,0847	0,0974	0,1353	0,1942	0,3059	0,58
0,1175	0,1282	0,1656	0,2060	0,2728	0,4001	0,0847	0,0887	0,1021	0,1422	0,2038	0,3185	0,59
0,1200	0,1303	0,1684	0,2096	0,2768	0,4065	0,0865	0,0907	0,1046	0,1447	0,2088	0,3250	0,595
0,1214	0,1325	0,1715	0,2133	0,2811	0,4130	0,0883	0,0926	0,1070	0,1493	0,2138	0,3315	0,6
0,1976	0,1623	0,1452	0,1638	0,2093	0,2630	0,1057	0,0934	0,1013	0,1178	0,1460	0,1795	0
0,1846	0,1524	0,1383	0,1584	0,2083	0,2797	0,1043	0,0928	0,1015	0,1204	0,1544	0,2068	0,1
0,1619	0,1344	0,1238	0,1438	0,1948	0,2827	0,0966	0,0865	0,0953	0,1152	0,1534	0,2270	0,2
0,1344	0,1127	0,1068	0,1257	0,1760	0,2787	0,0856	0,0777	0,0861	0,1064	0,1475	0,2442	0,3
0,1019	0,0874	0,0877	0,1048	0,1519	0,2672	0,0700	0,0658	0,0731	0,0931	0,1352	0,2558	0,4
0,0769	0,0668	0,0701	0,0838	0,1230	0,2282	0,0554	0,0535	0,0591	0,0767	0,1141	0,2338	0,45
0,0540	0,0469	0,0491	0,0591	0,0881	0,1710	0,0401	0,0388	0,0430	0,0562	0,0850	0,1835	0,5
0,0282	0,0243	0,0257	0,0310	0,0471	0,0958	0,0218	0,0210	0,0233	0,0309	0,0474	0,1078	0,55
0,0173	0,0149	0,0158	0,0189	0,0290	0,0601	0,0136	0,0130	0,0145	0,0192	0,0297	0,0690	0,57
0,0116	0,0101	0,0105	0,0128	0,0196	0,0410	0,0092	0,0088	0,0099	0,0130	0,0203	0,0475	0,58
0,0059	0,0051	0,0053	0,0065	0,0099	0,0210	0,0047	0,0045	0,0050	0,0066	0,0103	0,0246	0,59
0,0029	0,0026	0,0027	0,0032	0,0050	0,0106	0,0024	0,0023	0,0025	0,0033	0,0052	0,0124	0,595
0	0	0	0	0	0	0	0	0	0	0	0	0,6
0	0	0	0	0	0	0	0	0	0	0	0	0
0,0085	0,0102	0,0114	0,0232	0,0536	0,1983	0,0045	0,0051	0,0086	0,0151	0,0402	0,1350	0,1
0,0172	0,0212	0,0236	0,0471	0,1056	0,3590	0,0099	0,0112	0,0193	0,0335	0,0869	0,2706	0,2
0,0268	0,0330	0,0360	0,0712	0,1545	0,4901	0,0165	0,0187	0,0326	0,0556	0,1422	0,4173	0,3
0,0367	0,0452	0,0482	0,0945	0,1983	0,5939	0,0246	0,0274	0,0489	0,0823	0,2082	0,5843	0,4
0,0410	0,0506	0,0529	0,1034	0,2121	0,6135	0,0287	0,0317	0,0573	0,0955	0,2412	0,6591	0,45
0,0594	0,0679	0,0726	0,1263	0,2421	0,6184	0,0427	0,0464	0,0737	0,1177	0,2671	0,6700	0,5
0,0783	0,0857	0,0929	0,1493	0,2729	0,6297	0,0583	0,0628	0,0920	0,1427	0,2985	0,6727	0,55
0,0860	0,0929	0,1009	0,1587	0,2848	0,6360	0,0650	0,0699	0,1000	0,1536	0,3120	0,6844	0,57
0,0899	0,0966	0,1051	0,1633	0,2909	0,6395	0,0685	0,0736	0,1042	0,1593	0,3194	0,6914	0,58
0,0939	0,1002	0,1092	0,1680	0,2970	0,6433	0,0720	0,0773	0,1084	0,1650	0,3270	0,6992	0,59
0,0958	0,1021	0,1112	0,1704	0,3001	0,6453	0,0739	0,0792	0,1105	0,1681	0,3310	0,7034	0,595
0,0979	0,1040	0,1134	0,1728	0,3034	0,6473	0,0757	0,0812	0,1127	0,1711	0,3351	0,7078	0,6
0,1798	0,3109	0,5715	1,3762	2,4185	2,2960	0,0975	0,1245	0,2117	0,3903	0,9471	1,6087	0
0,1669	0,2855	0,5238	1,2735	2,2976	2,3345	0,0959	0,1223	0,2083	0,3882	0,9748	1,8195	0,1
0,1463	0,2460	0,4478	1,0941	2,0212	2,2077	0,0889	0,1134	0,1922	0,3604	0,9322	1,9085	0,2
0,1225	0,1966	0,3479	0,8321	1,5581	1,8459	0,0794	0,1006	0,1657	0,3063	0,7945	1,7589	0,3
0,0954	0,1363	0,2189	0,4662	0,8488	1,1234	0,0666	0,0830	0,1381	0,2329	0,5350	1,1960	0,4
0,0748	0,0942	0,1324	0,2212	0,3498	0,5127	0,0543	0,0668	0,0910	0,1390	0,2621	0,5136	0,45
0,0526	0,0666	0,0939	0,1586	0,2548	0,3895	0,0395	0,0488	0,0662	0,1034	0,1987	0,4085	0,5
0,0277	0,0351	0,0499	0,0850	0,1389	0,2215	0,0215	0,0267	0,0369	0,0575	0,1129	0,2429	0,55
0,0168	0,0214	0,0306	0,0525	0,0863	0,1399	0,0134	0,0166	0,0230	0,0361	0,0712	0,1574	0,57
0,0113	0,0144	0,0207	0,0354	0,0585	0,0959	0,0090	0,0113	0,0156	0,0245	0,0488	0,1088	0,58
0,0057	0,0073	0,0106	0,0182	0,0298	0,0491	0,0046	0,0058	0,0080	0,0125	0,0250	0,0563	0,59
0,0029	0,0038	0,0052	0,0091	0,0150	0,0246	0,0023	0,0029	0,0040	0,0063	0,0126	0,0286	0,595
0	0	0	0	0	0	0	0	0	0	0	0	0,6

$\tau^*=0,6$; funct. VIII; $\psi=45°$ $\qquad\qquad I^{(1)}(\tau,\ \theta,\ \psi)$

$\zeta,°$	30						45°					
$\theta,°$ / τ	0	15	30	45	60	75	0	15	30	45	60	75
0	0	0	0	0	0	0	0	0	0	0	0	0
0,1	0,0121	0,0105	0,0143	0,0232	0,0311	0,0994	0,0104	0,0077	0,0159	0,0193	0,0370	0,1121
0,2	0,0245	0,0209	0,0284	0,0457	0,0592	0,1725	0,0211	0,0155	0,0320	0,0382	0,0708	0,1958
0,3	0,0368	0,0313	0,0423	0,0674	0,0833	0,2243	0,0321	0,0231	0,0483	0,0563	0,1007	0,2564
0,4	0,0489	0,0411	0,0555	0,0875	0,1032	0,2584	0,0431	0,0304	0,0644	0,0731	0,1255	0,2965
0,45	0,0541	0,0449	0,0607	0,0949	0,1068	0,2540	0,0477	0,0329	0,0710	0,0786	0,1313	0,2929
0,5	0,0807	0,0684	0,0836	0,1189	0,1384	0,2905	0,0689	0,0540	0,0912	0,1023	0,1612	0,3256
9,55	0,1072	0,0916	0,1061	0,1419	0,1685	0,3223	0,0901	0,0749	0,1111	0,1254	0,1902	0,3555
0,57	0,1177	0,1008	0,1149	0,1509	0,1794	0,3341	0,0986	0,0834	0,1192	0,1346	0,2006	0,3670
0,58	0,1230	0,1054	0,1193	0,1554	0,1849	0,3397	0,1029	0,0875	0,1232	0,1390	0,2061	0,3725
0,59	0,1282	0,1100	0,1238	0,1598	0,1906	0,3452	0,1079	0,0918	0,1271	0,1436	0,2116	0,3780
0,595	0,1309	0,1123	0,1260	0,1621	0,1933	0,3479	0,1093	0,0939	0,1292	0,1459	0,2143	0,3808
0,6	0,1335	0,1146	0,1282	0,1643	0,1960	0,3505	0,1115	0,0960	0,1311	0,1482	0,2170	0,3833

$I^{(2)}(\tau,\ \theta,\ \psi)$

τ	0	15	30	45	60	75	0	15	30	45	60	75
0	0,6542	1,0262	1,1061	0,8742	0,7603	0,6225	0,3581	0,5066	0,6675	0,7763	0,8049	0,7727
0,1	0,5704	0,8934	0,9656	0,7713	0,6876	0,5955	0,3191	0,4496	0,5937	0,6964	0,7365	0,7484
0,2	0,4655	0,7296	0,7870	0,6342	0,5808	0,5404	0,2669	0,3736	0,4926	0,5832	0,6460	0,6828
0,3	0,3432	0,5279	0,5733	0,4746	0,4516	0,4655	0,2071	0,2835	0,3718	0,4441	0,5061	0,5799
0,4	0,2056	0,2919	0,3208	0,2866	0,2984	0,3696	0,1391	0,1785	0,2273	0,2775	0,3295	0,4326
0,45	0,1212	0,1510	0,1663	0,1669	0,1889	0,2735	0,0946	0,1115	0,1361	0,1681	0,1971	0,2959
0,5	0,0846	0,1061	0,1162	0,1171	0,1339	0,2019	0,0659	0,0781	0,0954	0,1139	0,1406	0,2200
0,55	0,0437	0,0543	0,0606	0,0615	0,0709	0,1134	0,0342	0,0405	0,0501	0,0600	0,0749	0,1223
0,57	0,0262	0,0328	0,0369	0,0376	0,0435	0,0695	0,0207	0,0247	0,0305	0,0367	0,0462	0,0766
0,58	0,0178	0,0221	0,0249	0,0255	0,0294	0,0473	0,0141	0,0167	0,0207	0,0247	0,0310	0,0522
0,59	0,0083	0,0105	0,0125	0,0129	0,0150	0,0243	0,0070	0,0083	0,0105	0,0124	0,0159	0,0266
0,595	0,0047	0,0057	0,0062	0,0004	0,0075	0,0121	0,0037	0,0043	0,0052	0,0063	0,0079	0,0134
0,6	0	0	0	0	0	0	0	0	0	0	0	0

$\tau^*=0,6$; funct. VIII; $\psi=90°$ $\qquad\qquad I^{(1)}(\tau,\ \theta,\ \psi)$

τ	0	15	30	45	60	75	0	15	30	45	60	75
0	0	0	0	0	0	0	0	0	0	0	0	0
0,1	0,0121	0,0136	0,0150	0,0164	0,0313	0,0687	0,0104	0,0105	0,0114	0,0168	0,0306	0,0627
0,2	0,0245	0,0273	0,0300	0,0324	0,0601	0,1197	0,0211	0,0214	0,0230	0,0336	0,0593	0,1105
0,3	0,0368	0,0409	0,0448	0,0475	0,0858	0,1567	0,0321	0,0323	0,0346	0,0500	0,0858	0,1461
0,4	0,0489	0,0543	0,0591	0,0613	0,1084	0,1820	0,0431	0,0431	0,0460	0,0655	0,1094	0,1711
0,45	0,0541	0,0600	0,0649	0,0659	0,1143	0,1779	0,0477	0,0475	0,0501	0,0709	0,1162	0,1670
0,5	0,0807	0,0860	0,0902	0,0942	0,1470	0,2238	0,0689	0,0691	0,0734	0,0970	0,1467	0,2133
0,55	0,1072	0,1117	0,1152	0,1213	0,1781	0,2636	0,0901	0,0907	0,0963	0,1124	0,1762	0,2539
0,57	0,1177	0,1220	0,1250	0,1320	0,1894	0,2781	0,0986	0,0994	0,1056	0,1325	0,1871	0,2690
0,58	0,1230	0,1271	0,1299	0,1374	0,1952	0,2850	0,1029	0,1036	0,1102	0,1374	0,1927	0,2762
0,59	0,1282	0,1322	0,1348	0,1426	0,2010	0,2915	0,1072	0,1080	0,1147	0,1425	0,1982	0,2833
0,595	0,1309	0,1349	0,1373	0,1452	0,2038	0,2949	0,1093	0,1102	0,1170	0,1451	0,2009	0,2867
0,6	0,1335	0,1374	0,1398	0,1478	0,2066	0,2982	0,1115	0,1124	0,1193	0,1474	0,2037	0,2901

$I^{(2)}(\tau,\ \theta,\ \psi)$

τ	0	15	30	45	60	75	0	15	30	45	60	75
0	0,6542	0,5918	0,4988	0,4071	0,3485	0,3808	0,3581	0,3472	0,3163	0,2849	0,2951	0,3369
0,1	0,5704	0,5160	0,4385	0,3639	0,3206	0,3711	0,3191	0,3100	0 2848	0,2613	0,2774	0,3356
0,2	0,4655	0,4234	0,3628	0,3059	0,2790	0,3455	0,2669	0,2602	0,2414	0,2263	0,2464	0,3191
0,3	0,3432	0,3146	0,2744	0,2419	0,2331	0,3115	0,2071	0,2031	0,1925	0,1874	0,2101	0,2935
0,4	0,2056	0,1926	0,1776	0,1716	0,1844	0,2687	0,1391	0,1384	0,1374	0,1451	0,1689	0,2590
0,45	0,1212	0,1180	0,1153	0,1223	0,1410	0,2125	0,0946	0,0956	0,0996	0,1121	0,1308	0,2073
0,5	0,0846	0,0827	0,0803	0,0854	0,0997	0,1565	0,0659	0,0669	0,0695	0,0759	0,0929	0,1536
0,55	0,0437	0,0423	0,0417	0,0447	0,0525	0,0879	0,0342	0,0346	0,0364	0,0397	0,0491	0,0848
0,57	0,0262	0,0255	0,0253	0,0273	0,0322	0,0536	0,0207	0,0211	0,0221	0,0242	0,0301	0,0529
0,58	0,0178	0,0172	0,0170	0,0184	0,0217	0,0364	0,0141	0,0142	0,0150	0,0162	0,0202	0,0360
0,59	0,0083	0,0081	0,0086	0,0093	0,0111	0,0186	0,0070	0,0071	0,0076	0,0082	0,0103	0,0183
0,595	0,0047	0,0045	0,0042	0,0046	0,0055	0,0094	0,0037	0,0037	0,0038	0,0042	0,0052	0,0093
0,6	0	0	0	0	0	0	0	0	0	0	0	0

Table I 85

TABLE I (continued)

60						75						ζ,°
0	15	30	45	60	75	0	15	30	45	60	75	θ,° \ τ
0	0	0	0	0	0	0	0	0	0	0	0	0
0,0083	0,0104	0,0103	0,0182	0,0395	0,1031	0,0045	0,0047	0,0073	0,0117	0,0220	0,0642	0,1
0,0172	0,0215	0,0214	0,0369	0,0781	0,1849	0,0099	0,0103	0,0163	0,0260	0,0466	0,1252	0,2
0,0268	0,0335	0,0328	0,0558	0,1146	0,2485	0,0165	0,0171	0,0273	0,0428	0,0742	0,1860	0,3
0,0367	0,0461	0,0439	0,0740	0,1481	0,2941	0,0246	0,0250	0,0407	0,0630	0,1048	0,2481	0,4
0,0410	0,0517	0,0482	0,0807	0,1585	0,3120	0,0287	0,0289	0,0475	0,0728	0,1173	0,2656	0,45
0,0594	0,0692	0,0677	0,1036	0,1859	0,3310	0,0427	0,0436	0,0636	0,0924	0,1450	0,3006	0,5
0,0783	0,0871	0,0878	0,1266	0,2138	0,3678	0,0583	0,0598	0,0815	0,1141	0,1761	0,3412	0,55
0,0860	0,0944	0,0957	0,1359	0,2242	0,3825	0,0650	0,0668	0,0893	0,1235	0,1890	0,3595	0,57
0,0899	0,0981	0,0999	0,1405	0,2297	0,3897	0,0685	0,0705	0,0932	0,1284	0,1959	0,3691	0,58
0,0939	0,1018	0,1039	0,1453	0,2352	0,3970	0,0720	0,0742	0,0973	0,1334	0,2029	0,3790	0,59
0,0958	0,1037	0,1059	0,1477	0,2379	0,4006	0,0739	0,0761	0,0994	0,1359	0,2066	0,3841	0,595
0,0979	0,1056	0,1080	0,1500	0,2407	0,4042	0,0757	0,0781	0,1016	0,1386	0,2103	0,3892	0,6
0,1798	0,2624	0,3783	0,5107	0,6395	0,7325	0,0975	0,1124	0,1539	0,2279	0,3272	0,4107	0
0,1669	0,2418	0,3481	0,4734	0,6062	0,7383	0,0959	0,1105	0,1520	0,2272	0,3350	0,4545	0,1
0,1463	0,2095	0,2997	0,4095	0,5350	0,6973	0,0889	0,1025	0,1412	0,2121	0,3202	0,4703	0,2
0,1225	0,1701	0,2383	0,3247	0,4327	0,6075	0,0794	0,0914	0,1247	0,1856	0,2839	0,4503	0,3
0,0954	0,1225	0,1616	0,2161	0,2942	0,4526	0,0666	0,0765	0,1140	0,1616	0,2452	0,4163	0,4
0,0748	0,0886	0,1083	0,1388	0,1879	0,3003	0,0543	0,0623	0,0796	0,1059	0,1527	0,2595	0,45
0,0526	0,0626	0,0766	0,0989	0,1358	0,2266	0,0395	0,0455	0,0577	0,0784	0,1147	0,2038	0,5
0,0277	0,0330	0,0405	0,0526	0,0733	0,1276	0,0215	0,0248	0,0321	0,0433	0,0645	0,1192	0,55
0,0168	0,0202	0,0248	0,0324	0,0454	0,0804	0,0134	0,0154	0,0200	0,0271	0,0405	0,0774	0,57
0,0113	0,0135	0,0168	0,0219	0,0307	0,0550	0,0090	0,0105	0,0136	0,0184	0,0277	0,0534	0,58
0,0057	0,0069	0,0086	0,0112	0,0156	0,0281	0,0046	0,0053	0,0069	0,0094	0,0142	0,0285	0,59
0,0029	0,0035	0,0042	0,0056	0,0079	0,0141	0,0023	0,0027	0,0035	0,0047	0,0072	0,0140	0,595
0	0	0	0	0	0	0	0	0	0	0	0	0,6
0	0	0	0	0	0	0	0	0	0	0	0	0
0,0083	0,0089	0,0111	0,0154	0,0240	0,0501	0,0045	0,0048	0,0054	0,0072	0,0119	0,0273	0,1
0,0172	0,0186	0,0231	0,0316	0,0478	0,0905	0,0099	0,0105	0,0120	0,0158	0,0249	0,0518	0,2
0,0268	0,0289	0,0358	0,0486	0,0707	0,1229	0,0165	0,0175	0,0199	0,0255	0,0391	0,0747	0,3
0,0367	0,0397	0,0488	0,0655	0,0923	0,1477	0,0246	0,0259	0,0293	0,0368	0,0542	0,0962	0,4
0,0410	0,0444	0,0544	0,0722	0,0986	0,1451	0,0287	0,0302	0,0338	0,0416	0,0593	0,0974	0,45
0,0594	0,0632	0,0745	0,0951	0,1278	0,1915	0,0427	0,0446	0,0497	0,0607	0,0846	0,1385	0,5
0,0783	0,0823	0,0950	0,1182	0,1569	0,2336	0,0583	0,0605	0,0672	0,0815	0,1122	0,1801	0,55
0,0860	0,0902	0,1033	0,1275	0,1677	0,2496	0,0650	0,0675	0,0748	0,0904	0,1234	0,1972	0,57
0,0899	0,0942	0,1075	0,1321	0,1734	0,2574	0,0685	0,0710	0,0787	0,0950	0,1293	0,2059	0,58
0,0939	0,0981	0,1118	0,1368	0,1790	0,2651	0,0720	0,0747	0,0827	0,0997	0,1353	0,2147	0,59
0,0958	0,1002	0,1138	0,1392	0,1818	0,2688	0,0739	0,0766	0,0847	0,1021	0,1384	0,2191	0,595
0,0979	0,1022	0,1160	0,1416	0,1847	0,2726	0,0757	0,0785	0,0868	0,1045	0,1415	0,2236	0,6
0,1798	0,1800	0,1844	0,2012	0,2337	0,2613	0,0975	0,0995	0,1079	0,1226	0,1421	0,1538	0
0,1669	0,1673	0,1723	0,1905	0,2277	0,2714	0,0959	0,0980	0,1071	0,1233	0,1476	0,1723	0,1
0,1463	0,1469	0,1520	0,1698	0,2088	0,2682	0,0889	0,0909	0,0999	0,1165	0,1440	0,1833	0,2
0,1225	0,1234	0,1287	0,1454	0,1837	0,2579	0,0794	0,0812	0,0897	0,1060	0,1357	0,1906	0,3
0,0954	0,0968	0,1022	0,1169	0,1520	0,2386	0,0666	0,0682	0,0885	0,1086	0,1488	0,2379	0,4
0,0748	0,0761	0,0808	0,0919	0,1192	0,1970	0,0543	0,0555	0,0616	0,0738	0,0995	0,1659	0,45
0,0526	0,0537	0,0569	0,0651	0,0856	0,1477	0,0395	0,0404	0,0443	0,0542	0,0741	0,1290	0,5
0,0277	0,0282	0,0300	0,0344	0,0458	0,0826	0,0215	0,0219	0,0245	0,0297	0,0413	0,0744	0,55
0,0168	0,0172	0,0183	0,0211	0,0283	0,0518	0,0134	0,0136	0,0152	0,0185	0,0258	0,0485	0,57
0,0113	0,0115	0,0123	0,0142	0,0190	0,0354	0,0090	0,0093	0,0103	0,0125	0,0176	0,0333	0,58
0,0057	0,0058	0,0063	0,0073	0,0096	0,0180	0,0046	0,0047	0,0053	0,0064	0,0090	0,0172	0,59
0,0029	0,0030	0,0031	0,0036	0,0049	0,0090	0,0023	0,0023	0,0026	0,0032	0,0045	0,0087	0,595
0	0	0	0	0	0	0	0	0	0	0	0	0

τ* = 0,6; **funct.** VIII; ψ = 135° $I^{(1)}(\tau, \theta, \psi)$

ζ,°	30						45					
θ,°	0	15	30	45	60	75	0	15	30	45	60	75
τ												
0	0	0	0	0	0	0	0	0	0	0	0	0
0,1	0,0121	0,0132	0,0109	0,0182	0,0317	0,0523	0,0104	0,0150	0,0127	0,0175	0,0304	0,0481
0,2	0,0245	0,0264	0,0216	0,0357	0,0611	0,0912	0,0211	0,0306	0,0256	0,0349	0,0591	0,0852
0,3	0,0368	0,0397	0,0323	0,0527	0,0877	0,1190	0,0321	0,0466	0,0387	0,0524	0,0862	0,1132
0,4	0,0489	0,0527	0,0421	0,0685	0,1114	0,1377	0,0431	0,0630	0,0518	0,0693	0,1113	0,1333
0,45	0,0541	0,0582	0,0456	0,0743	0,1183	0,1322	0,0477	0,0705	0,0570	0,0756	0,1192	0,1577
0,5	0,0807	0,0880	0,0794	0,1104	0,1566	0,1930	0,0689	0,0943	0,0866	0,1101	0,1588	0,1915
0,55	0,1072	0,1175	0,1128	0,1455	0,1932	0,2453	0,0901	0,1182	0,1163	0,1442	0,1975	0,2465
0,57	0,1177	0,1294	0,1260	0,1595	0,2067	0,2642	0,0986	0,1279	0,1281	0,1575	0,2119	0,2669
0,58	0,1230	0,1352	0,1326	0,1663	0,2136	0,2736	0,1029	0,1326	0,1341	0,1645	0,2191	0,2766
0,59	0,1282	0,1412	0,1393	0,1732	0,2204	0,2823	0,1072	0,1376	0,1400	0,1714	0,2264	0,2861
0,595	0,1309	0,1442	0,1426	0,1767	0,2237	0,2867	0,1093	0,1400	0,1430	0,1748	0,2301	0,2908
0,6	0,1335	0,1470	0,1459	0,1799	0,2271	0,2908	0,1115	0,1424	0,1459	0,1780	0,2337	0,2953

$I^{(2)}(\tau, \theta, \psi)$

τ	0	15	30	45	60	75	0	15	30	45	60	75
0	0,6542	0,4355	0,3134	0,2575	0,2540	0,2775	0,3581	0,2332	0,2020	0,2001	0,1891	0,2631
0,1	0,5704	0,3809	0,2801	0,2325	0,2366	0,2748	0,3191	0,2100	0,1838	0,1859	0,1823	0,2677
0,2	0,4655	0,3142	0,2343	0,1985	0,2100	0,2626	0,2669	0,1788	0,1581	0,1636	0,1500	0,2612
0,3	0,3432	0,2382	0,1851	0,1622	0,1812	0,2485	0,2071	0,1453	0,1306	0,1394	0,1365	0,2505
0,4	0,2056	0,1551	0,1330	0,1260	0,1517	0,2339	0,1391	0,1095	0,1015	0,1133	0,1258	0,2349
0,45	0,1212	0,1031	0,0973	0,0973	0,1210	0,1993	0,0946	0,0843	0,0792	0,0908	0,1153	0,1980
0,5	0,0846	0,0722	0,0676	0,0677	0,0853	0,1469	0,0618	0,0589	0,0551	0,0613	0,0818	0,1468
0,55	0,0437	0,0368	0,0350	0,0353	0,0448	0,0826	0,0342	0,0304	0,0286	0,0319	0,0432	0,0811
0,57	0,0262	0,0222	0,0212	0,0216	0,0275	0,0502	0,0207	0,0185	0,0174	0,0195	0,0265	0,0506
0,58	0,0178	0,0150	0,0143	0,0145	0,0185	0,0341	0,0141	0,0125	0,0118	0,0131	0,0178	0,0346
0,59	0,0083	0,0071	0,0072	0,0073	0,0095	0,0174	0,0070	0,0063	0,0059	0,0065	0,0091	0,0175
0,595	0,0047	0,0039	0,0035	0,0036	0,0047	0,0088	0,0037	0,0032	0,0029	0,0033	0,0045	0,0089
0,6	0	0	0	0	0	0	0	0	0	0	0	0

τ*=0,6 **funct.** VIII; ψ=180° $I^{(1)}(\tau, \theta, \psi)$

τ	0	15	30	45	60	75	0	15	30	45	60	75
0	0	0	0	0	0	0	0	0	0	0	0	0
0,1	0,0121	0,0152	0,0178	0,0155	0,0321	0,0536	0,0104	0,0151	0,0112	0,0231	0,0288	0,0549
0,2	0,0245	0,0304	0,0357	0,0314	0,0617	0,0935	0,0211	0,0310	0,0225	0,0461	0,0561	0,0970
0,3	0,0368	0,0457	0,0537	0,0448	0,0885	0,1223	0,0321	0,0473	0,0341	0,0692	0,0813	0,1284
0,4	0,0489	0,0607	0,0711	0,0580	0,1127	0,1419	0,0431	0,0640	0,0452	0,0920	0,1043	0,1510
0,45	0,0541	0,0673	0,0786	0,0624	0,1198	0,1368	0,0477	0,0715	0,0495	0,1015	0,1111	0,1467
0,5	0,0807	0,0969	0,1084	0,1029	0,1669	0,2021	0,0689	0,0976	0,0823	0,1357	0,1621	0,2268
0,55	0,1072	0,1260	0,1377	0,1421	0,2118	0,2584	0,0901	0,1238	0,1150	0,1691	0,2116	0,2974
0,57	0,1177	0,1377	0,1492	0,1577	0,2288	0,2288	0,0986	0,1345	0,1282	0,1825	0,2303	0,3236
0,58	0,1230	0,1436	0,1551	0,1654	0,2371	0,2888	0,1029	0,1397	0,1349	0,1892	0,2397	0,3364
0,59	0,1282	0,1493	0,1609	0,1731	0,2456	0,2982	0,1072	0,1451	0,1414	0,1961	0,2491	0,3487
0,595	0,1309	0,1523	0,1639	0,1769	0,2497	0,3029	0,1093	0,1478	0,1447	0,1993	0,2539	0,3548
0,6	0,1335	0,1552	0,1668	0,1807	0,2539	0,3077	0,1115	0,1505	0,1479	0,2027	0,2586	0,3607

$I^{(2)}(\tau, \theta, \psi)$

τ	0	15	30	45	60	75	0	15	30	45	60	75
0	0,6542	0,3899	0,2703	0,2520	0,2254	0,2968	0,3581	0,2003	0,1955	0,1847	0,2101	0,2759
0,1	0,5704	0,3414	0,2430	0,2265	0,2098	0,2930	0,3191	0,1808	0,1773	0,1706	0,1991	0,2776
0,2	0,4655	0,2822	0,2020	0,1922	0,1866	0,2783	0,2669	0,1549	0,1519	0,1496	0,1547	0,2686
0,3	0,3432	0,2160	0,1623	0,1555	0,1635	0,2579	0,2071	0,1279	0,1250	0,1282	0,1364	0,2567
0,4	0,2056	0,1444	0,1201	0,1200	0,1429	0,2382	0,1391	0,1004	0,0964	0,1074	0,1234	0,2444
0,45	0,1213	0,0990	0,0905	0,0915	0,1181	0,1992	0,0946	0,0799	0,0745	0,0882	0,1147	0,2104
0,5	0,0846	0,0693	0,0629	0,0638	0,0832	0,1468	0,0659	0,0557	0,0519	0,0597	0,0813	0,1562
0,55	0,0437	0,0354	0,0326	0,0332	0,0439	0,0826	0,0342	0,0288	0,0270	0,0309	0,0429	0,0865
0,57	0,0262	0,0213	0,0197	0,0203	0,0269	0,0502	0,0207	0,0175	0,0163	0,0190	0,0264	0,0540
0,58	0,0178	0,0144	0,0132	0,0136	0,0182	0,0342	0,0141	0,0119	0,0110	0,0127	0,0177	0,0368
0,59	0,0083	0,0068	0,0067	0,0069	0,0092	0,0175	0,0070	0,0059	0,0056	0,0063	0,0089	0,0188
0,595	0,0047	0,0037	0,0033	0,0034	0,0046	0,0087	0,0037	0,0031	0,0028	0,0032	0,0045	0,0094
0	0	0	0	0	0	0	0	0	0	0	0	0

Table I 87

TABLE I (continued)

| 60 | | | | | | 75 | | | | | | ζ° |
0	15	30	45	60	75	0	15	30	45	60	75	θ,° / τ
0	0	0	0	0	0	0	0	0	0	0	0	0
0,0083	0,0079	0,0114	0,0158	0,0230	0,0480	0,0045	0,0046	0,0043	0,0059	0,0114	0,0271	0,1
0,0172	0,0165	0,0237	0,0325	0,0460	0,0872	0,0099	0,0102	0,0093	0,0128	0,0237	0,0515	0,2
0,0268	0,0255	0,0369	0,0500	0,0684	0,1189	0,0165	0,0170	0,0151	0,0204	0,0372	0,0743	0,3
0,0367	0,0348	0,0508	0,0682	0,0900	0,1439	0,0246	0,0250	0,0218	0,0289	0,0518	0,0959	0,4
0,0410	0,0387	0,0570	0,0757	0,0966	0,1422	0,0287	0,0291	0,0246	0,0322	0,0570	0,0974	0,45
0,0594	0,0589	0,0802	0,1046	0,1351	0,2011	0,0427	0,0442	0,0425	0,0546	0,0875	0,1480	0,5
0,0783	0,0794	0,1039	0,1339	0,1735	0,2545	0,0583	0,0609	0,0623	0,0791	0,1208	0,1992	0,55
0,0860	0,0879	0,1134	0,1458	0,1881	0,2750	0,0650	0,0682	0,0708	0,0897	0,1344	0,2203	0,57
0,0899	0,0922	0,1183	0,1517	0,1957	0,2849	0,0685	0,0719	0,0752	0,0951	0,1416	0,2311	0,58
0,0939	0,0965	0,1233	0,1578	0,2032	0,2947	0,0720	0,0758	0,0797	0,1006	0,1490	0,2420	0,59
0,0958	0,0986	0,1257	0,1608	0,2070	0,2996	0,0739	0,0777	0,0820	0,1034	0,1527	0,2475	0,595
0,0979	0,1008	0,1282	0,1639	0,2108	0,3044	0,0757	0,0798	0,0843	0,1063	0,1565	9,2530	0,6

| 60 | | | | | | 75 | | | | | | ζ° |
0	15	30	45	60	75	0	15	30	45	60	75	θ,° / τ
0,1798	0,1518	0,1413	0,1406	0,1910	0,2295	0,0975	0,0877	0,0898	0,1085	0,1287	0,1453	0
0,1669	0,1416	0,1332	0,1350	0,1873	0,2391	0,0959	0,0866	0,0896	0,1097	0,1344	0,1628	0,1
0,1463	0,1248	0,1188	0,1225	0,1733	0,2384	0,0889	0,0809	0,0843	0,1044	0,1323	0,1739	0,2
0,1225	0,1057	0,1026	0,1094	0,1558	0,2345	0,0794	0,0733	0,0772	0,0963	0,1270	0,1844	0,3
0,0954	0,0843	0,0849	0,0960	0,1355	0,2292	0,0666	0,0634	0,0681	0,0826	0,1459	0,2411	0,4
0,0748	0,0670	0,0693	0,0811	0,1112	0,1998	0,0543	0,0531	0,0579	0,0706	0,1013	0,1778	0,45
0,0526	0,0471	0,0486	0,0573	0,0798	0,1500	0,0395	0,0386	0,0415	0,0518	0,0754	0,1386	0,5
0,0277	0,0247	0,0255	0,0302	0,0427	0,0840	0,0215	0,0210	0,0230	0,0284	0,0421	0,0801	0,55
0,0168	0,0150	0,0156	0,0185	0,0263	0,0527	0,0134	0,0131	0,0142	0,0177	0,0263	0,0522	0,57
0,0113	0,0101	0,0105	0,0125	0,0177	0,0360	0,0090	0,0089	0,0097	0,0120	0,0180	0,0359	0,58
0,0057	0,0051	0,0054	0,0064	0,0089	0,0183	0,0046	0,0045	0,0049	0,0061	0,0092	0,0185	0,59
0,0029	0,0026	0,0027	0,0032	0,0045	0,0092	0,0023	0,0022	0,0025	0,0031	0,0046	0,0093	0,595
0	0	0	0	0	0	0	0	0	0	0	0	0,6

| 60 | | | | | | 75 | | | | | | ζ° |
0	15	30	45	60	75	0	15	30	45	60	75	θ,° / τ
0	0	0	0	0	0	0	0	0	0	0	0	0
0,0083	0,0087	0,0116	0,0143	0,0270	0,0428	0,0045	0,0043	0,0043	0,0062	0,0105	0,0288	0,1
0,0172	0,0181	0,0243	0,0294	0,0543	5,0780	0,0099	0,0093	0,0094	0,0133	0,0219	0,0559	0,2
0,0268	0,0280	0,0379	0,0454	0,0816	0,1071	0,0165	0,0154	0,0155	0,0214	0,0343	0,0833	0,3
0,0367	0,0385	0,0523	0,0616	0,1090	0,1308	0,0246	0,0225	0,0225	0,0307	0,0477	0,1125	0,4
0,0410	0,0429	0,0587	0,0682	0,1189	0,1293	0,0287	0,0259	0,0256	0,0346	0,0522	0,1199	0,45
0,0594	0,0637	0,0865	0,1047	0,1619	0,2142	0,0427	0,0415	0,0447	0,0626	0,0934	0,1801	0,5
0,0783	0,0849	0,1152	0,1417	0,2051	0,2916	0,0583	9,0588	0,0658	0,0936	0,1385	0,2419	0,55
0,0860	0,0937	0,1268	0,1566	0,2216	0,3212	0,0650	0,0663	0,0749	0,1070	0,1572	0,2677	0,57
0,0899	0,0981	0,1326	0,1642	0,2300	0,3357	0,0685	0,0702	0,0796	0,1139	0,1670	0,2809	0,58
0,0939	0,1026	0,1385	0,1718	0,2387	0,3499	0,0720	0,0743	0,0844	0,1209	0,1770	0,2942	0,59
0,0958	0,1048	0,1415	0,1757	0,2429	0,3570	0,0739	0,0762	0,0868	0,1245	0,1822	0,3010	0,595
0,0979	0,1070	0,1446	0,1795	0,2472	0,3640	0,0757	0,0784	0,0894	0,1282	0,1874	0,3077	0,6

| 60 | | | | | | 75 | | | | | | ζ° |
0	15	30	45	60	75	0	15	30	45	60	75	θ,° / τ
0,1798	0,1461	0,1213	0,1352	0,1677	0,2013	0,0975	0,0830	0,0923	0,1052	0,1253	0,1467	0
0,1669	0,1365	0,1153	0,1308	0,1683	0,2164	0,0959	0,0821	0,0922	0,1072	0,1329	0,1704	0,1
0,1463	0,1204	0,1040	0,1202	0,1604	0,2243	0,0889	0,0770	0,0868	0,1033	0,1335	0,1908	0,2
0,1225	0,1022	0,0922	0,1084	0,1504	0,2322	0,0794	0,0702	0,0793	0,0970	0,1317	0,2135	0,3
0,0954	0,0815	0,0800	0,0955	0,0384	0,2410	0,0666	0,0617	0,0690	0,1057	0,1545	0,2844	0,4
0,0748	0,0647	0,0677	0,0808	0,1189	0,2204	0,0543	0,0525	0,0579	0,0753	0,1121	0,2288	0,45
0,0526	0,1455	0,0475	0,0571	0,0853	0,1658	0,0395	0,0381	0,0415	0,0553	0,0836	0,1795	0,5
0,0277	0,0238	0,0250	0,0301	0,0457	0,0931	0,0215	0,0207	0,0230	0,0304	0,0468	0,1048	0,55
0,0168	0,0145	0,0153	0,0185	0,0282	0,0585	0,0134	0,0128	0,0143	0,0190	0,0293	0,0682	0,57
0,0113	0,0098	0,0102	0,0124	0,0191	0,0400	0,0090	0,0088	0,0097	0,0128	0,0200	0,0470	0,58
0,0057	0,0050	0,0052	0,0064	0,0096	0,0204	0,0046	0,0044	0,0049	0,0065	0,0103	0,0242	0,59
0,0029	0,0025	0,0026	0,0032	0,0049	0,0102	0,0023	0,0022	0,0025	0,0033	0,0052	0,0123	0,595
0	0	0	0	0	0	0	0	0	0	0	0	0,6

$\tau^* = 0,8$; **funct.** VII; $\psi = 0°$ $I^{(1)}(\tau, \theta, \psi)$

$\varsigma,°$	30						45					
$\theta,°$ / τ	0	15	30	45	60	75	0	15	30	45	60	75
0	0	0	0	0	0	0	0	0	0	0	0	0
0,2	0,0404	0,0363	0,0479	0,0635	0,0941	0,2271	0,0340	0,0311	0,0436	0,0583	0,1041	0,2545
0,4	0,0820	0,0735	0,0961	0,1239	0,1725	0,3618	0,0702	0,0640	0,0890	0,1154	0,1935	0,4113
0,5	0,1022	0,0909	0,1186	0,1511	0,2037	0,4028	0,0882	0,0798	0,1108	0,1417	0,2280	0,4654
0,6	0,1218	0,1076	0,1400	0,1758	0,2291	0,4290	0,1061	0,0950	0,1320	0,2841	0,2639	0,5022
0,7	0,1727	0,1512	0,1814	0,2185	0,2798	0,4652	0,1467	0,1342	0,1693	0,3106	0,3093	0,5356
0,75	0,1977	0,1725	0,2012	0,2385	0,3029	0,4808	0,1669	0,1536	0,1876	0,3241	0,3314	0,5538
0,78	0,2127	0,1852	0,2131	0,2501	0,3160	0,4896	0,1790	0,1652	0,1986	0,3323	0,3441	0,5644
0,79	0,2177	0,1893	0,2170	0,2539	0,3203	0,4925	0,1831	0,1691	0,2021	0,3350	0,3483	0,5681
0,795	0,2202	0,1914	0,2189	0,2557	0,3226	0,4939	0,1852	0,1710	0,2039	0,3364	0,3505	0,5700
0,798	0,2217	0,1926	0,2200	0,2569	0,3236	0,4946	0,1863	0,1720	0,2051	0,3371	0,3517	0,5711
0,8	0,2227	0,1934	0,2207	0,2577	0,3246	0,4952	0,1872	0,1729	0,2058	0,3378	0,3526	0,5718

$I^{(2)}(\tau, \theta, \psi)$

τ	0	15	30	45	60	75	0	15	30	45	60	75
0	0,6854	1,2946	1,8749	1,5986	1,0920	0,8730	0,4136	0,6390	1,2299	1,8981	1,7470	1,1411
0,2	0,5654	1,0806	1,5781	1,3691	0,9676	0,7093	0,3494	0,5430	1,0616	1,6812	1,5886	1,0029
0,4	0,3928	0,7331	1,0743	0,9587	0,7190	0,7286	0,2552	0,3899	0,7523	1,2023	1,1987	1,0022
0,5	0,2877	0,5097	0,7266	0,6698	0,5377	0,6003	0,1974	0,2886	0,5303	0,8325	0,8658	0,8133
0,6	0,1661	0,2333	0,2984	0,3102	0,3102	0,4132	0,1303	0,1675	0,2518	0,3496	0,4087	0,4960
0,7	0,0896	0,1282	0,1637	0,1713	0,1745	0,2504	0,0705	0,0917	0,1390	0,1962	0,2348	0,3068
0,75	0,0461	0,0656	0,0853	0,0903	0,0922	0,1377	0,0366	0,0475	0,0730	0,1036	0,1255	0,1705
0,78	0,0187	0,0266	0,0350	0,0372	0,0381	0,0581	0,0150	0,0194	0,0302	0,0428	0,0521	0,0724
0,79	0,0088	0,0127	0,0177	0,0189	0,0194	0,0296	0,0073	0,0097	0,0153	0,0216	0,0266	0,0370
0,795	0,0050	0,0069	0,0089	0,0095	0,0096	0,0148	0,0039	0,0049	0,0077	0,0109	0,0134	0,0185
0,798	0,0039	0,0065	0,0036	0,0068	0,0032	0,0056	0,0015	0,0018	0,0056	0,0043	9,0054	0,0075
0,8	0	0	0	0	0	0	0	0	0	0	0	0

$\tau^* = 0,8$; **funct.** VII; $\psi = 45°$ $I^{(1)}(\tau, \theta, \psi)$

τ	0	15	30	45	60	75	0	15	30	45	60	75
0	0	0	0	0	0	0	0	0	0	0	0	0
0,2	0,0404	0,0387	0,0460	0,0627	0,0883	0,1992	0,0340	0,0315	0,0431	0,0544	0,0889	0,2026
0,4	0,0820	0,0783	0,0922	0,1225	0,1628	0,3185	0,0702	0,0647	0,0879	0,1074	0,1661	0,3290
0,5	0,1022	0,0971	0,1136	0,1496	0,1925	0,3550	0,0882	0,0807	0,1098	0,1319	0,1962	0,3726
0,6	0,1218	0,1151	0,1339	0,1744	0,2170	0,3786	0,1061	0,0963	0,1310	0,2729	0,2267	0,4025
0,7	0,1727	0,1596	0,1770	0,2179	0,2682	0,4248	0,1467	0,1361	0,1693	0,3003	0,2765	0,4452
0,75	0,1977	0,1813	0,1978	0,2385	0,2914	0,4440	0,1669	0,1559	0,1882	0,3138	0,2995	0,4642
0,78	0,2127	0,1944	0,2102	0,2505	0,3047	0,4543	0,1790	0,1677	0,1994	0,3220	0,3127	0,4749
0,79	0,2177	0,1987	0,2142	0,2544	0,3091	0,4577	0,1831	0,1716	0,2030	0,3247	0,3171	0,4785
0,795	0,2202	0,2008	0,2162	0,2563	0,3113	0,4594	0,1852	0,1736	0,2049	0,3261	0,3192	0,4804
0,798	0,2217	0,2021	0,2174	0,2575	0,3124	0,4603	0,1863	0,1747	0,2061	0,3268	0,3206	0,4813
0,8	0,2227	0,2029	0,2182	0,2584	0,3134	0,4610	0,1872	0,1756	0,2068	0,3275	0,3215	0,4820

$I^{(2)}(\tau, \theta, \psi)$

τ	0	15	30	45	60	75	0	15	30	45	60	75
0	0,6854	0,9789	1,0435	0,8720	0,7806	0,6539	0,4136	0,5380	0,6711	0,7533	0,7984	0,7037
0,2	0,5654	0,8144	0,8736	0,7395	0,6900	0,5155	0,3494	0,4562	0,5744	0,6578	0,6994	0,5889
0,4	0,3928	0,5583	0,6066	0,5283	0,5216	0,5578	0,2552	0,3300	0,4160	0,4844	0,5409	0,6264
0,5	0,2877	0,3980	0,4323	0,3907	0,4033	0,4781	0,1974	0,2482	0,3108	0,3647	0,4203	0,5374
0,6	0,1661	0,2032	0,2243	0,2272	0,2584	0,3670	0,1303	0,1518	0,1839	0,2177	0,2652	0,3882
0,7	0,0896	0,1113	0,1221	0,1244	0,1445	0,2215	0,0705	0,0828	0,1007	0,1205	0,1502	0,2380
0,75	0,0461	0,0568	0,0634	0,0652	0,0762	0,1215	0,0366	0,0429	0,0526	0,0632	0,0797	0,1316
0,78	0,0187	0,0229	0,0260	0,0268	0,0314	0,0513	0,0150	0,0175	0,0217	0,0260	0,0329	0,0557
0,79	0,0088	0,0110	0,0130	0,0136	0,0159	0,0261	0,0073	0,0088	0,0110	0,0130	0,0168	0,0285
0,795	0,0050	0,0060	0,0067	0,0068	0,0080	0,0131	0,0039	0,0044	0,0055	0,0066	0,0084	0,0143
0,798	0,0039	0,0056	0,0026	0,0048	0,0026	0,0050	0,0015	0,0016	0,0040	0,0026	0,0035	0,0058
0,8	0	0	0	0	0	0	0	0	0	0	0	0

Table I 89

TABLE I (continued)

60						75						ζ,°
0	15	30	45	60	75	0	15	30	45	60	75	θ,° / τ
0	0	0	0	0	0	0	0	0	0	0	0	0
0,0252	0,0278	0,0327	0,0523	0,1010	0,2726	0,0115	0,0127	0,0176	0,0275	0,0558	0,1640	0,2
0,0545	0,0600	0,0697	0,1091	0,2000	0,4778	0,0279	0,0307	0,0428	0,0662	0,1313	0,3124	0,4
0,0693	0,0769	0,0882	0,1370	0,2460	0,5626	0,0375	0,0413	0,0584	0,0899	0,1786	0,4035	0,5
0,0850	0,0940	0,1066	0,1642	0,2891	0,6363	0,0491	0,0538	0,0772	0,1187	0,2367	0,5218	0,6
0,1201	0,1271	0,1433	0,2049	0,3381	0,6471	0,0752	0,0811	0,1073	0,1583	0,2852	0,5556	0,7
0,1383	0,1442	0,1620	0,2258	0,3636	0,6613	0,0902	0,0969	0,1251	0,1822	0,3166	0,5930	0,75
0,1494	0,1546	0,1734	0,2384	0,3792	0,6721	0,1001	0,1073	0,1367	0,1979	0,3379	0,6224	0,78
0,1531	0,1582	0,1773	0,2427	0,3844	0,6762	0,1037	0,1111	0,1409	0,2035	0,3457	0,6335	0,79
0,1550	0,1600	0,1792	0,2448	0,3872	0,6783	0,1054	0,1127	0,1430	0,2063	0,3496	0,6393	0,795
0,1561	0,1610	0,1804	0,2460	0,3887	0,6796	0,1064	0,1138	0,1442	0,2080	0,3519	0,6429	0,798
0,1569	0,1617	0,1811	0,2469	0,3899	0,6804	0,1072	0,1148	0,1451	0,2092	0,3535	0,6452	0,8

60						75						ζ,°
0	15	30	45	60	75	0	15	30	45	60	75	θ,° / τ
0,2328	0,3412	0,5440	1,0987	1,7304	1,5026	0,1155	0,1417	0,2050	0,3204	0,6135	0,8117	0
0,2060	0,3037	0,4911	1,0262	1,7126	1,5437	0,1105	0,1373	0,2038	0,3322	0,6966	1,0429	0,2
0,1630	0,2349	0,3761	0,7925	1,3704	1,5722	0,0974	0,1215	0,1807	0,3003	0,6645	1,2655	0,4
0,1350	0,1865	0,2881	0,5799	1,0013	1,2503	0,0860	0,1066	0,1548	0,2526	0,5484	1,0896	0,5
0,1022	0,1266	0,1747	0,2856	0,4428	0,6271	0,0703	0,0855	0,1146	0,1721	0,3150	0,5436	0,6
0,0562	0,0704	0,0982	0,1641	0,2625	0,4036	0,0413	0,0508	0,0691	0,1061	0,2025	0,4166	0,7
0,0294	0,0369	0,0519	0,0877	0,1428	0,2288	0,0224	0,0276	0,0379	0,0589	0,1148	0,2480	0,75
0,0121	0,0151	0,0214	0,0364	0,0599	0;0987	0,0094	0,0116	0,0160	0,0250	0,0496	0,1101	0,78
0,0061	0,0077	0,0109	0,0186	0,0304	0,0505	0,0048	0,0059	0,0082	0,0128	0,0253	0,0569	0,79
0,0031	0,0039	0,0054	0,0093	0,0153	0,0253	0,0024	0,0029	0,0041	0,0063	0,0128	0,0290	0,795
0,0012	0,0015	0,0018	0,0038	0,0061	0,0103	0,0009	0,0012	0,0016	0,0026	0,0052	0,0116	0,798
0	0	0	0	0	0	0	0	0	0	0	0	0,8

60						75						ζ,°
0	15	30	45	60	75	0	15	30	45	60	75	θ,° / τ
0	0	0	0	0	0	0	0	0	0	0	0	0
0,0252	0,0278	0,0310	0,0452	0,0803	0,1745	0,0115	0,0123	0,0160	0,0223	0,0391	0,1074	0,2
0,0545	0,0601	0,0662	0,0942	0,1594	0,3019	0,0279	0,0295	0,0388	0,0553	0,0885	0,1803	0,4
0,0693	0,0770	0,0837	0,1182	0,1957	0,3498	0,0375	0,0397	0,0524	0,0742	0,1166	0,2173	0,5
0,0850	0,0944	0,1011	0,1413	0,2296	0,3878	0,0491	0,0516	0,0687	0,0968	0,1489	0,2631	0,6
0,1201	0,1279	0,1375	0,1824	0,2755	0,4382	0,0752	0,0787	0,0983	0,1320	0,1961	0,3269	0,7
0,1383	0,1452	0,1561	0,2032	0,2986	0,4642	0,0902	0,0943	0,1157	0,1527	0,2242	0,3669	0,75
0,1494	0,1557	0,1673	0,2159	0,3127	0,4801	0,1001	0,1047	0,1269	0,1662	0,2427	0,3941	0,78
0,1531	0,1593	0,1712	0,2200	0,3173	0,4855	0,1037	0,1084	0,1310	0,1710	0,2493	0,4038	0,79
0,1550	0,1611	0,1730	0,2221	0,3197	0,4882	0,1054	0,1100	0,1330	0,1734	0,2527	0,4087	0,795
0,1561	0,1622	0,1742	0,2234	0,3211	0,4899	0,1064	0,1111	0,1342	0,1748	0,2546	0,4118	0,798
0,1569	0,1629	0,1749	0,2243	0,3221	0,4909	0,1072	0,1121	0,1350	0,1759	0,2560	0,4138	0,8

60						75						ζ,°
0	15	30	45	60	75	0	15	30	45	60	75	θ,° / τ
0,2328	0,3028	0,3966	0,5004	0,5874	0,6014	0,1155	0,1313	0,1654	0,2165	0,2726	0,2781	0
0,2060	0,2695	0,3565	0,4603	0,5673	0,5500	0,1105	0,1267	0,1631	0,2209	0,2975	0,2972	0,2
0,1630	0,2102	0,2766	0,3657	0,4676	0,6217	0,0974	0,1121	0,1459	0,2015	0,2865	0,3919	0,4
0,1350	0,1694	0,2182	0,2868	0,3748	0,5349	0,0860	0,0987	0,1278	0,1752	0,2521	0,3599	0,5
0,1022	0,1197	0,1449	0,1844	0,2467	0,3810	0,0703	0,0801	0,1011	0,1331	0,1885	0,2641	0,6
0,0562	0,0664	0,0809	0,1043	0,1434	0,2406	0,0413	0,0474	0,0606	0,0811	0,1186	0,2123	0,7
0,0294	0,0348	0,0426	0,0553	0,0772	0,1350	0,0224	0,0258	0,0331	0,0447	0,0664	0,1245	0,75
0,0121	0,0143	0,0175	0,0228	0,0321	0,0578	0,0094	0,0108	0,0140	0,0189	0,0285	0,0548	0,78
0,0061	0,0072	0,0090	0,0117	0,0163	0,0295	0,0048	0,0055	0,0071	0,0097	0,0145	0,0282	0,79
0,0031	0,0037	0,0044	0,0058	0,0081	0,0148	0,0024	0,0027	0,0036	0,0048	0,0073	0,0143	0,795
0,0012	0,0014	0,0015	0,0024	0,0033	0,0060	0,0009	0,0011	0,0014	0,0020	0,0030	0,0057	0,798
0	0	0	0	0	0	0	0	0	0	0	0	0,8

$\tau^* = 0,8$; funct. VII; $\psi = 90°$ $I^{(1)} (\tau, \theta, \psi)$

$\zeta,°$	30						45					
$\theta/°$ τ	0	15	30	45	60	75	0	15	30	45	60	75
0	0	0	0	0	0	0	0	0	0	0	0	0
0,2	0,0404	0,0427	0 0473	0,0558	0,0861	0,1604	0,0340	0,0350	0,0388	0,0507	0,0780	0,1425
0,4	0,0820	0,0867	0,0950	0,1093	0,1604	0,2595	0,0702	0,0720	0,0792	0,1011	0,1476	0,2335
0,5	0,1022	0,1077	0,1175	0 1334	0,1912	0,2899	0,0882	0,0902	0,0988	0,1247	0,1780	0,2650
0,6	0,1218	0,1283	0,1389	0,1554	0,2175	0,3105	0,1061	0,1083	0,1176	0,2650	0,2048	0,2872
0,7	0,1727	0,1777	0,1865	0,2059	0,2719	0,3718	0,1467	0,1493	0,1609	0,2966	0,2567	0,3506
0,75	0,1977	0,2021	0,2097	0,2300	0,2968	0,3966	0,1669	0,1698	0,1823	0,3124	0,2808	0,3771
0,78	0,2127	0,2167	0,2234	0,2440	0,3110	0,4097	0,1790	0,1821	0,1951	0,3219	0,2947	0,3916
0,79	0,2177	0,2215	0,2280	0,2485	0,3156	0,4139	0,1831	0,1861	0,1992	0,3251	0,2993	0,3963
0,795	0,2202	0,2239	0,2301	0,2509	0,3179	0,4160	0,1852	0,1882	0,2014	0,3268	0,3016	0,3987
0,798	0,2217	0,2253	0,2314	0,2522	0,3193	0,4173	0,1863	0,1893	0,2026	0,3276	0,3030	0,4000
0,8	0,2227	0,2262	0,2323	0,2532	0,3203	0,4180	0,1872	0,1902	0,2035	0,3283	0,3040	0,4009

$I^{(2)} (\tau, \theta, \psi)$

τ	0	15	30	45	60	75	0	15	30	45	60	75
0	0,6854	0,6357	0,5602	0,4946	0,4491	0,4480	0,4136	0,4062	0,3861	0,3650	0,4034	0,3835
0,2	0,5654	0,5251	0,4674	0,4194	0,3974	0,3467	0,3494	0,3442	0,3299	0,3186	0,3403	0,3683
0,4	0,3928	0,3663	0,3335	0,3114	0,3148	0,3974	0,2552	0,2539	0,2474	0,2469	0,2786	0,3587
0,5	0,2877	0,2722	0,2523	0,2454	0,2607	0,3535	0,1974	0,1969	0,1964	0,2023	0,2357	0,3331
0,6	0,1661	0,1616	0,1600	0,1706	0,1987	0,2938	0,1303	0,1319	0,1378	0,1509	0,1832	0,2828
0,7	0,0896	0,0879	0,0862	0,0927	0,1102	0,1759	0,0705	0,0716	0,0749	0,0824	0,1023	0,1713
0,75	0,0461	0,0447	0,0445	0,0483	0,0579	0,0960	0,0366	0,0370	0,0389	0,0430	0,0538	0,0940
0,78	0,0187	0,0180	0,0181	0,0198	0,0238	0,0405	0,0150	0,0151	0,0160	0,0176	0,0222	0,0395
0,79	0,0088	0,0086	0,0091	0,0100	0,0120	0,0206	0,0073	0,0075	0,0080	0,0088	0,0112	0,0203
0,795	0,0050	0,0047	0,0046	0,0050	0,0060	0,0103	0,0039	0,0038	0,0041	0,0044	0,0057	0,0101
0,798	0,0039	0,0043	0,0019	0,0035	0,0020	0,0040	0,0015	0,0014	0,0029	0,0017	0,0023	0,0041
0,8	0	0	0	0	0	0	0	0	0	0	0	0

$\tau^* = 0,8$; funct. VII; $\psi = 135°$ $I^{(1)} (\tau, \theta, \psi)$

τ	0	15	30	45	60	75	0	15	30	45	60	75
0	0	0	0	0	0	0	0	0	0	0	0	0
0,2	0,0404	0,0432	0,0444	0,0589	0,0879	0,1437	0,0340	0,0402	0,0413	0,0531	0,0797	0,1296
0,4	0,0820	0,0877	0,0891	0,1159	0,1644	0,2337	0,0702	0,0639	0,0844	0.1064	0,1519	0,2138
0,5	0,1022	0,1092	0,1099	0,1418	0,1967	0,2611	0,0882	0,1048	0,1056	0,1317	0,1867	0,2433
0,6	0,1218	0,1301	0,1290	0,1660	0,2249	0,2799	0,1061	0,1266	0,1262	0,2742	0,2134	0,2647
0,7	0,1727	0,1865	0,1924	0,2316	0,2895	0,3633	0,1467	0,1723	0,1815	0,3210	0,2807	0,3518
0,75	0,1977	0,2143	0,2230	0,2630	0,3192	0,3970	0,1669	0,1546	0,2091	0,3449	0,3128	0,3891
0,78	0,2127	0,2311	0,2413	0,2814	0,3363	0,4150	0,1790	0,2091	0,2257	0,3592	0,3316	0,4094
0,79	0,2177	0,2365	0,2474	0,2874	0,3419	0,4208	0,1831	0,2136	0,2311	0,3640	0,3377	0,4159
0,795	0,2202	0,2393	0,2503	0,2906	0,3446	0,4236	0,1852	0,2160	0,2340	0,3665	0,3409	0,4192
0,798	0,2217	0,2405	0,2521	0,2923	0,3462	0,4253	0,1863	0,2173	0,2356	0,3678	0,3426	0,4210
0,8	0,2227	0,2420	0,2533	0,2936	0,3475	0,4263	0,1872	0,2183	0,2367	0,3688	0,3439	0,4222

$I^{(2)} (\tau, \theta, \psi)$

τ	0	15	30	45	60	75	0	15	30	45	60	75
0	0,6854	0,5040	0,4064	0,3554	0,3525	0,3631	0,4136	0,3111	0,2787	0,2775	0,3128	0,3258
0,2	0,5654	0,4153	0,3374	0,3004	0,3132	0,2783	0,3494	0,2634	0,2373	0,2419	0,2585	0,2625
0,4	0,3928	0,2935	0,2467	0,2273	0,2538	0,3365	0,2552	0,1988	0,1811	0,1904	0,2190	0,3165
0,5	0,2877	0,2233	0,1945	0,1851	0,2159	0,3108	0,1974	0,1600	0,1484	0,1596	0,1933	0,3033
0,6	0,1661	0,1428	0,1370	0,1393	0,1735	0,2780	0,1303	0,1176	0,1123	0,1251	0,1639	0,2716
0,7	0,0896	0,0773	0,0734	0,0750	0,0957	0,1663	0,0705	0,0636	0,0604	0,0678	0,0912	0,1646
0,75	0,0461	0,0393	0,0379	0,0389	0,0500	0,0907	0,0366	0,0328	0,0313	0,0352	0,0478	0,0902
0,78	0,0187	0,0158	0,0154	0,0160	0,0206	0,0382	0,0150	0,0133	0,0128	0,0143	0,0197	0,0381
0,79	0,0088	0,0077	0,0077	0,0081	0,0104	0,0194	0,0073	0,0067	0,0064	0,0072	0,0100	0,0195
0,795	0,0050	0,0042	0,0040	0,0040	0,0052	0,0097	0,0039	0,0034	0,0033	0,0036	0,0050	0,0097
0,798	0,0039	0,0038	0,0016	0,0027	0,0018	0,0037	0,0015	0,0012	0,0022	0,0014	0,0020	0,0039
0,8	0	0	0	0	0	0	0	0	0	0	0	0

Table I
91

TABLE I (continued)

		60							75				ζ,°
													θ,°
0	15	30	45	60	75	0	15	30	45	60	75		τ
0	0	0	0	0	0	0	0	0	0	0	0		0
0,0252	0,0264	0,0309	0,0402	0,0592	0,1094	0,0115	0,0121	0,0140	0,0182	0,0276	0,0720		0,2
0,0545	0,0572	0,0664	0,0845	0,1183	0,1911	0,0279	0,0292	0,0332	0,0421	0,0607	0,1040		0,4
0,0693	0,0732	0,0844	0,1064	0,1451	0,2207	0,0375	0,0393	0,0445	0,0555	0,0782	0,1153		0,5
0,0850	0,0896	0,1028	0,1280	0,1703	0,2444	0,0491	0,0510	0,0575	0,0708	0,0971	0,1296		0,6
0,1201	0,1253	0,1407	0,1698	0,2197	0,3096	0,0752	0,0780	0,0867	0,1048	0,1401	0,1972		0,7
0,1383	0,1438	0,1601	0,1910	0,2440	0,3390	0,0902	0,0935	0 1036	0,1243	0,1644	0,2328		0,75
0,1494	0,1551	0,1719	0,2039	0,2584	0,3557	0,1001	0,1037	0,1146	0,1370	0,1800	0,2552		0,78
0,1531	0,1589	0,1759	0,2080	0,2632	0,3613	0,1037	0,1075	0,1185	0,1414	0,1855	0,2629		0,79
0,1550	0,1608	0,1779	0,2102	0,2657	0,3641	0,1054	0,1090	0,1204	0,1436	0,1883	0,2668		0,795
0,1561	0,1620	0,1791	0,2115	0,2671	0,3657	0,1064	0,1101	0,1216	0,1450	0,1899	0,2692		0,798
0,1569	0,1627	0,1799	0,2124	0,2681	0,3668	0,1072	0,1111	0,1224	0,1459	0,1911	0,2708		0,8

0	15	30	45	60	75	0	15	30	45	60	75	τ
0,2328	0,2338	0,2293	0,2547	0,2794	0,2842	0,1155	0,1176	0,1251	0,1374	0,1507	0,1406	0
0,2060	0,2075	0,2143	0,2333	0,2696	0,2457	0,1105	0,1128	0,1218	0,1375	0,1603	),1636	0,2
0,1630	0,1647	0,1720	0,1919	0,2349	0,3125	0,0974	0,0996	0,1090	0,1266	0,1579	),1889	0,4
0,1350	0,1369	0,1439	0,1622	0,2036	0,2941	0,0860	0,0880	0,0968	0,1139	0,1466	0,1827	0,5
0,1022	0,1041	0,1108	0,1265	0,1631	0,2605	0,0703	0,0720	0,0797	0,0951	0,1268	0,1606	0,6
0,0562	0,0574	0,0612	0,0705	0,0931	0,1617	0,0413	0,0423	0,0471	0,0569	0,0781	0,1378	0,7
0,0294	0,0300	0,0321	0,0370	0,0496	0,0899	0,0224	0,0230	0,0256	0,0311	0,0432	0,0799	0,75
0,0121	0,0123	0,0131	0,0153	0,0205	0,0382	0,0094	0,0096	0,0107	0,0130	0,0184	0,0348	0,78
0,0061	0,0062	0,0067	0,0078	0,0104	0,0195	0,0048	0,0049	0,0055	0,0067	0,0094	0,0179	0,79
0,0031	0,0032	0,0033	0,0038	0,0052	0,0098	0,0023	0,0024	0,0027	0,0033	0,0047	0,0091	0,795
0,0012	0,0013	0,0011	0,0016	0,0021	0,0040	0,0009	0,0010	0,0011	0,0014	0,0019	0,0036	0,798
0	0	0	0	0	0	0	0	0	0	0	0	0,8

0	15	30	45	60	75	0	15	30	45	60	75	τ
0	0	0	0	0	0	0	0	0	0	0	0	0
0,0252	0,0258	0,0317	0,0414	0,0588	0,1061	0,0115	0,0120	0,0130	0,0171	0,0268	0,0707	0,2
0,0545	0,0557	0,0684	0,0873	0,1184	0,1872	0,0279	0,0287	0,0306	0,0393	0,0592	0,1022	0,4
0,0693	0,0711	0,0873	0,1104	0,1458	0,2172	0,0375	0,0387	0,0407	0,0515	0,0765	0,1136	0,5
0,0850	0,0868	0,1069	0,1338	0,1722	0,2423	0,0491	0,0503	0,0521	0,0653	0,0955	0,1284	0,6
0,1201	0,1248	0,1503	0,1865	0,2380	0,3292	0,0752	0,0784	0,0846	0,1049	0,1477	0,2125	0,7
0,1383	0,1445	0,1726	0,2135	0,2707	0,3686	0,0902	0,0944	0,1035	0,1278	0,1774	0,2567	0,75
0,1494	0,1565	0,1864	0,2300	0,2902	0,3909	0,1001	0,1047	0,1157	0,1426	0,1965	0,2847	0,78
0,1531	0,1606	0,1910	0,2354	0,2967	0,3984	0,1037	0,1085	0,1201	0,1479	0,2033	0,2944	0,79
0,1550	0,1627	0,1934	0,2382	0,3000	0,4021	0,1054	0,1100	0,1223	0,1504	0,2067	0,2993	0,795
0,1561	0,1639	0,1948	0,2399	0,3020	0,4044	0,1064	0,1111	0,1236	0,1520	0,2087	0,3023	0,798
0,1569	0,1647	0,1956	0,2410	0,3033	0,4058	0,1072	0,1121	0,1245	0,1532	0,2101	0,3496	0,8

0	15	30	45	60	75	0	15	30	45	60	75	τ
0,2328	0,2026	0,1924	0,1965	0,2345	0,2510	0,1155	0,1078	0,1103	0,1241	0,1388	0,1322	0
0,2060	0,1794	0,1723	0,1805	0,2273	0,2160	0,1105	0,1034	0,1072	0,1245	0,1486	0,1553	0,2
0,1630	0,1429	0,1404	0,1492	0,2029	0,2871	0,0974	0,0920	0,0971	0,1162	0,1495	0,1832	0,4
0,1350	0,1196	0,1199	0,1319	0,1809	0,2789	0,0860	0,0822	0,0878	0,1061	0,1420	0,1832	0,5
0,1022	0,0928	0,0964	0,1131	0,1533	0,2636	0,0703	0,0691	0,0752	0,0913	0,1286	0,1737	0,6
0,0562	0,0508	0,0530	0,0628	0,0873	0,1641	0,0413	0,0405	0,0443	0,0546	0,0794	0,1473	0,7
0,0294	0,0265	0,0276	0,0328	0,0465	0,0913	0,0224	0,0220	0,0240	0,0298	0,0439	0,0856	0,75
0,0121	0,0108	0,0113	0,0135	0,0192	0,0389	0,0094	0,0092	0,0101	0,0125	0,0187	0,0374	0,78
0,0061	0,0055	0,0057	0,0069	0,0097	0,0199	0,0048	0,0046	0,0051	0,0064	0,0095	0,0192	0,79
0,0031	0,0028	0,0028	0,0034	0,0049	0,0100	0,0024	0,0023	0,0026	0,0031	0,0048	0,0098	0,795
0,0012	0,0011	0,0010	0,0014	0,0019	0,0040	0,0009	0,0009	0,0010	0,0013	0,0019	0,0039	0,798
0	0	0	0	0	0	0	0	0	0	0	0	0,8

$\tau^* = 0,8$; funct.VII; $\psi = 180°$ $I^{(1)}(\tau, \theta, \psi)$

$\zeta,°$	30						45					
$\theta,°$ / τ	0	15	30	45	60	75	0	15	30	45	60	75
0	0	0	0	0	0	0	0	0	0	0	0	0
0,2	0,0404	0,0456	0,0520	0,0577	0,0898	0,1453	0,0340	0,0408	0,0405	0,0595	0,0795	0,1343
0,4	0,0820	0,0925	0,1046	0,1134	0.1684	0,2373	0,0702	0,0845	0,0829	0,1193	0.1517	0,2234
0,5	0,1022	0,1154	0,1298	0,1386	0,2018	0,2660	0.C882	0,1066	0,1035	0,1484	0.1876	0,2552
0,6	0,1218	0,1377	0,1540	0,1619	0,2312	0,2863	0,1061	0,1288	0,1237	0,2948	0,2131	0,2796
0,7	0,1727	0,1935	0,2098	0,2343	0,3111	0,3776	0,1467	0,1786	0.1846	0,3418	0.3016	0,4011
0,75	0,1977	0.2210	0,2370	0,2692	0,3483	0,4145	0,1669	0,2038	0,2151	0,3656	0,3431	0,4516
0,78	0,2127	0,2377	0,2533	0,2896	0,3696	0,4344	0,1790	0,2189	0,2332	0,3799	0.3673	0,4795
0,79	0,2177	0,2431	0,2587	0,2965	0,3765	0,4408	0,1831	0,2239	0,2394	0,3846	0,3752	0,4887
0,795	0,2202	0,2458	0,2613	0,2998	0,3801	0.4438	0,1852	0,2265	0,2424	0,3871	0.3794	0,4931
0,798	0,2217	0,2474	0,2629	0,3019	0,3821	0,4456	0,1863	0,2279	0,2442	0,3885	0.3818	0,4957
0,8	0,2227	0,2485	0,2640	0,3033	0,3836	0,4468	0,1872	0,2290	0,2455	0,3895	0,3834	0,4975

$I^{(2)}(\tau, \theta, \psi)$

τ	0	15	30	45	60	75	0	15	30	45	60	75
0	0,6854	0,4643	0,3642	0,3315	0,3168	0,3606	0,4136	0,2814	0,2613	0,2486	0,3000	0,3091
0,2	0,5654	0,3824	0,3003	0,2809	0,2827	0,2825	0,3494	0,2380	0,2227	0,2183	0,2498	0,2545
0,4	0,3928	0,2719	0,2212	0,2139	0,2340	0,3392	0,2552	0,1811	0,1708	0,1755	0,2153	0,3140
0,5	0,2877	0,2090	0,1767	0,1749	0,2034	0,3135	0,1974	0,1476	0,1404	0,1502	0,1916	0,3085
0,6	0,1661	0,1376	0,1284	0,1325	0,1701	0,2781	0,1303	0,1120	0.1067	0,1225	0,1634	0,2869
0,7	0,0896	0,0744	0,0687	0,0710	0,0938	0,1665	0,0705	0,0604	0,0572	0,0663	0,0910	0,1744
0,75	0,0461	0,0379	0,0354	0,0368	0,0489	0,0908	0,0366	0,0312	0,0297	0,0342	0,0477	0,0958
0,78	0,0187	0,0152	0,0144	0,0151	0,0201	0,0382	0,0150	0,0126	0,0120	0,0140	0,0196	0,0406
0,79	0,0088	0,0073	0,0073	0,0076	0,0102	0,0194	0,0073	0,0064	0,0060	0,0070	0,0099	0,0206
0,795	0,0050	0,0040	0,0037	0,0038	0,0051	0,0098	0,0039	0,0032	0,0031	0,0035	0,0050	0,0102
0,798	0,0039	0,0036	0,0015	0,0025	0,0017	0,0038	0,0015	0,0011	0,0021	0,0014	0,0020	0,0042
0,8	0	0	0	0	0	0	0	0	0	0	0	0

$\tau^* = 0,8$; funct. VIII; $\psi = 0°$ $I^{(1)}(\tau, \theta, \psi)$

τ	0	15	30	45	60	75	0	15	30	45	60	75
0	0	0	0	0	0	0	0	0	0	0	0	0
0,2	0,0249	0,0192	0,0327	0,0470	0,0698	0,2073	0,0210	0,0168	0,0313	0,0430	0,0874	0,2445
0,4	0,0503	0,0386	0,0655	0,0915	0,1265	0,3246	0,0435	0,0343	0,0638	0,0845	0,1611	0,3872
0,5	0,0628	0,0473	0,0809	0,1115	0,1480	0,3585	0,0547	0,0422	0,0793	0,1030	0,1911	0,4297
0,6	0,0748	0,0554	0,0957	0,1298	0,1642	0,3767	0,0656	0,0494	0,0944	0,1193	0,2142	0,4516
0,7	0,1270	0,1010	0,1385	0,1740	0,2210	0,4198	0,1073	0,0902	0,1328	0,1642	0,2643	0,4937
0,75	0,1529	0,1235	0,1592	0,1951	0,2470	0,4386	0,1281	0,1106	0,1516	0,1860	0,2882	0,5146
0,78	0,1683	0,1367	0,1715	0,2073	0,2618	0,4490	0,1406	0,1227	0,1628	0,1987	0,3020	0,5269
0,79	0,1735	0,1410	0,1756	0,2114	0,2666	0,4523	0,1449	0,1268	0,1666	0,2029	0,3066	0,5312
0,795	0,1761	0,1433	0,1775	0,2134	0,2691	0,4540	0,1470	0,1288	0,1686	0,2051	0,3089	0,5332
0,798	0,1775	0,1445	0,1787	0,2146	0,2705	0,4549	0,1482	0,1300	0,1697	0,2063	0,3104	0,5345
0,8	0,1786	0,1455	0,1796	0,2154	0,2713	0,4557	0,1490	0,1308	0,1704	0,2072	0,3112	0,5354

$I^{(2)}(\tau, \theta, \psi)$

τ	0	15	30	45	60	75	0	15	30	45	60	75
0	0,7822	1,6370	2,4597	2,0333	1,2455	0,9692	0,4288	0,7250	1,5432	2,4738	2,1904	1,3565
0,2	0,6307	1,3446	2,0424	1,7129	1,1093	0,7413	0,3551	0,6046	1,3120	2,1596	1,9931	1,1460
0,4	0,4226	0,8816	1,3464	1,1571	0,7898	0,7433	0,2537	0,4199	0,9006	1,4995	1,4539	1,1200
0,5	0,3004	0,5885	0,8730	0,7748	0,5699	0,5765	0,1938	0,3024	0,6095	0,9959	1,0062	0,8683
0,6	0,1612	0,2282	0,2926	0,3032	0,3004	0,3917	0,1259	0,1630	0,2467	0,3432	0,4001	0,4811
0,7	0,0873	0,1259	0,1610	0,1681	0,1698	0,2418	0,0685	0,0896	0,1368	0,1933	0,2307	0,2996
0,75	0,0451	0,0645	0,0842	0,0887	0,0901	0,1334	0,0355	0,0316	0,0721	0,1023	0,1235	0,1665
0,78	0,0183	0,0262	0,0346	0,0366	0,0372	0,0565	0,0145	0,0190	0,0298	0,0424	0,0514	0,0709
0,79	0,0085	0,0125	0,0174	0,0186	0,0190	0,0287	0,0072	0,0095	0,0151	0,0213	0,0262	0,0363
0,795	0,0049	0,0068	0,0087	0,0092	0,0093	0,0203	0,0038	0,0048	0,0076	0,0108	0,0132	0,0181
0,798	0,0040	0,0064	0,0036	0,0067	0,0031	0,0055	0,0014	0,0018	0,0055	0,0043	0,0054	0,0074
0,8	0	0	0	0	0	0	0	0	0	0	0	0

Table I 93

TABLE I (continued)

	60						75					ξ,°
0	15	30	45	60	75	0	15	30	45	60	75	θ,° / τ
0	0	0	0	0	0	0	0	0	0	0	0	0
0,0252	0,0266	0,0326	0,0412	0,0641	0,1050	0,0115	0,0117	0,0132	0,0175	0,0268	0,0728	0,2
0,0545	0,0576	0,0705	0,0870	0,1299	0,1866	0,0279	0,0280	0,0311	0,0406	0,0595	0,1089	0,4
0,0693	0,0737	0,0901	0,1100	0,1612	0,2175	0,0375	0,0376	0,0414	0,0535	0,0770	0,1246	0,5
0,0850	0,0902	0,1104	0,1332	0,1921	0,2441	0,0491	0,0487	0,0533	0,0683	0,0965	0,1461	0,6
0,1201	0,1293	0,1622	0,1990	0,2666	0,3726	0,0752	0,0775	0,0878	0,1179	0,1664	0,2475	0,7
0,1383	0,1497	0,1891	0,2329	0,3039	0,4307	0,0902	0,0942	0,1080	0,1469	0,2067	0,3019	0,75
0,1494	0,1622	0,2056	0,2536	0,3263	0,4641	0,1001	0,1053	0,1211	0,1659	0,2329	0,3367	0,78
0,1531	0,1664	0,2113	0,2605	0,3338	0,4752	0,1037	0,1094	0,1258	0,1726	0,2421	0,3488	0,79
0,1550	0,1684	0,2141	0,2641	0,3377	0,4806	0,1054	0,1110	0,1281	0,1759	0,2468	0,3548	0,795
0,1561	0,1698	0,2158	0,2661	0,3398	0,4840	0,1064	0,1123	0,1295	0,1779	0,2496	0,3586	0,798
0,1569	0,1707	0,2169	0,2677	0,3415	0,4862	0,1072	0,1133	0,1305	0,1794	0,2515	0,3610	0,8

0	15	30	45	60	75	0	15	30	45	60	75	τ
0,2328	0,1949	0,1763	0,1902	0,2219	0,2398	0,1155	0,1044	0,1107	0,1235	0,1400	0,1370	0
0,2060	0,1726	0,1583	0,1761	0,2190	0,2152	0,1105	0,1001	0,1080	0,1249	0,1522	0,1406	0,2
0,1630	0,1378	0,1309	0,1456	0,2020	0,2928	0,0974	0,0894	0,0980	0,1186	0,1571	0,2131	0,4
0,1350	0,1156	0,1137	0,1300	0,1848	0,2931	0,0860	0,0803	0,0884	0,1098	0,1521	0,2252	0,5
0,1022	0,0899	0,0945	0,1127	0,1625	0,2878	0,0703	0,0683	0,0753	0,0968	0,1411	0,2298	0,6
0,0562	0,0492	0,0518	0,0625	0,0930	0,1802	0,0413	0,0401	0,0444	0,0581	0,0876	0,1882	0,7
0,0294	0,0257	0,0269	0,0327	0,0495	0,1006	0,0224	0,0217	0,0240	0,0317	0,0487	0,1103	0,75
0,0121	0,0104	0,0110	0,0134	0,0205	0,0429	0,0094	0,0090	0,0101	0,0133	0,0208	0,0485	0,78
0,0061	0,0053	0,0056	0,0068	0,0103	0,0219	0,0048	0,0046	0,0051	0,0068	0,0106	0,0250	0,79
0,0031	0,0027	0,0028	0,0034	0,0051	0,0109	0,0024	0,0023	0,0025	0,0034	0,0053	0,0127	0,795
0,0012	0,0011	0,0009	0,0014	0,0020	0,0044	0,0009	0,0009	0,0010	0,0014	0,0021	0,0051	0,798
0	0	0	0	0	0	0	0	0	0	0	0	0,8

0	15	30	45	60	75	0	15	30	45	60	75	τ
0	0	0	0	0	0	0	0	0	0	0	0	0
0,0161	0,0193	0,0224	0,0422	0,0922	0,2974	0,0077	0,0088	0,0139	0,0246	0,0566	0,1684	0,2
0,0350	0,0418	0,0475	0,0873	0,1795	0,5110	0,0190	0,0215	0,0345	0,0579	0,1338	0,3600	0,4
0,0447	0,0535	0,0596	0,1089	0,2182	0,5944	0,0257	0,0290	0,0473	0,0783	0,1816	0,4758	0,5
0,0546	0,0653	0,0710	0,1293	0,2519	0,6596	0,0338	0,0375	0,0628	0,1025	0,2401	0,6166	0,6
0,0905	0,0993	0,1092	0,1718	0,3039	0,6562	0,0603	0,0654	0,0934	0,1431	0,2864	0,6172	0,7
0,1092	0,1167	0,1288	0,1936	0,3308	0,6654	0,0757	0,0814	0,1114	0,1673	0,3167	0,6424	0,75
0,1206	0,1275	0,1406	0,2069	0,3472	0,6739	0,0857	0,0920	0,1232	0,1834	0,3376	0,6655	0,78
0,1244	0,1311	0,1447	0,2113	0,3527	0,6772	0,0892	0,0957	0,1274	0,1890	0,3452	0,6747	0,79
0,1264	0,1329	0,1467	0,2136	0,3556	0,6790	0,0910	0,0976	0,1296	0,1919	0,3490	0,6796	0,795
0,1276	0,1339	0,1479	0,2149	0,3573	0,6802	0,0921	0,0987	0,1308	0,1937	0,3513	0,6826	0,798
0,1284	0,1348	0,1487	0,2157	0,3584	0,6809	0,0928	0,0994	0,1317	0,1948	0,3530	0,6846	0,8

0	15	30	45	60	75	0	15	30	45	60	75	τ
0,2160	0,3513	0,6145	1,3839	2,2862	1,9687	0,1094	0,1347	0,2113	0,3583	0,6408	1,0768	0
0,1883	0,3078	0,5447	1,2636	2,2281	1,9683	0,1031	0,1288	0,2069	0,3654	0,8543	1,3673	0,2
0,1489	0,2337	0,4049	0,9449	1,7252	1,9440	0,0909	0,1158	0,1807	0,3215	0,7903	1,6246	0,4
0,1253	0,1838	0,3018	0,6648	1,2078	1,4792	0,0813	0,1011	0,1537	0,2641	0,6272	1,3563	0,5
0,0990	0,1232	0,1710	0,2816	0,4362	0,6158	0,0688	0,0838	0,1129	0,1698	0,3120	0,5849	0,6
0,0547	0,0688	0,0963	0,1618	0,2594	0,3978	0,0407	0,0499	0,0683	0,1050	0,2011	0,4146	0,7
0,0286	0,0361	0,0512	0,0866	0,1412	0,2259	0,0221	0,0272	0,0376	0,0583	0,1141	0,2467	0,75
0,0117	0,0149	0,0211	0,0360	0,0594	0,0975	0,0092	0,0115	0,0159	0,0248	0,0493	0,1096	0,78
0,0059	0,0075	0,0107	0,0184	0,0301	0,0499	0,0047	0,0058	0,0081	0,0127	0,0252	0,0567	0,79
0,0030	0,0038	0,0053	0,0091	0,0152	0,0251	0,0023	0,0030	0,0041	0,0064	0,0127	0,0288	0,795
0,0012	0,0015	0,0018	0,0037	0,0061	0,0101	0,0009	0,0011	0,0015	0,0026	0,0051	0,0116	0,798
0	0	0	0	0	0	0	0	0	0	0	0	0,8

τ*=0,8; **funct.**VIII; ψ=45° $I^{(1)}$ (τ, θ, ψ)

ζ,°	30						45					
θ,° τ	0	15	30	45	60	75	0	15	30	45	60	75
0	0	0	0	0	0	0	0	0	0	0	0	0
0,2	0,0249	0,0223	0,0296	0,0462	0,0634	0,1730	0,0210	0,0171	0,0307	0,0381	0,0695	0,1844
0,4	0,0503	0,0450	0,0591	0,0904	0,1157	0,2730	0,0435	0,0348	0,0630	0,0755	0,1293	0,2962
0,5	0,0628	0,0556	0,0728	0,1103	0,1353	0,3021	0,0547	0,0430	0,0785	0,0921	0,1535	0,3304
0,6	0,0748	0,0657	0,0857	0,1287	0,1515	0,3189	0,0656	0,0505	0,0938	0,1070	0,1727	0,3503
0,7	0,1270	0,1119	0,1307	0,1742	0,2089	0,3741	0,1073	0,0917	0,1330	0,1523	0,2264	0,3998
0,75	0,1529	0,1346	0,1524	0,1957	0,2350	0,3973	0,1281	0,1127	0,1525	0,1741	0,2517	0,4221
0,78	0,1683	0,1481	0,1653	0,2084	0,2501	0,4098	0,1406	0,1250	0,1639	0,1870	0,2662	0,4349
0,79	0,1735	0,1525	0,1696	0,2125	0,2549	0,4139	0,1449	0,1291	0,1678	0,1912	0,2710	0,4390
0,795	0,1761	0,1549	0,1717	0,2146	0,2574	0,4159	0,1470	0,1312	0,1697	0,1933	0,2734	0,4411
0,798	0,1775	0,1562	0,1730	0,2160	0,2588	0,4169	0,1482	0,1321	0,1709	0,1946	0,2749	0,4423
0,8	0,1786	0,1571	0,1737	0,2167	0,2597	0,4178	0,1490	0,1332	0,1713	0,1955	0,2758	0,4432

$I^{(2)}$ (τ, θ, ψ)

τ	0	15	30	45	60	75	0	15	30	45	60	75
0	0,7822	1,1885	1,2735	1,0171	0,8342	0,6823	0,4288	0,5907	0,7649	0,8740	0,8879	0,7894
0,2	0,6307	0,9698	1,0428	0,8404	0,7479	0,4921	0,3551	0,4908	0,6398	0,7420	0,7812	0,6130
0,4	0,4226	0,6410	0,6962	0,5741	0,5403	0,5285	0,2537	0,3438	0,4473	0,5238	0,5757	0,6468
0,5	0,3004	0,4402	0,4771	0,4101	0,4060	0,4468	0,1938	0,2531	0,3241	0,3811	0,4318	0,5319
0,6	0,1612	0,1981	0,2186	0,2202	0,2484	0,3453	0,1259	0,1473	0,1788	0,2113	0,2563	0,3729
0,7	0,0873	0,1090	0,1194	0,1210	0,1397	0,2127	0,0685	0,0807	0,0983	0,1176	0,1459	0,2304
0,75	0,0451	0,0557	0,0622	0,0636	0,0739	0,1171	0,0355	0,0419	0,0516	0,0618	0,0776	0,1274
0,78	0,0183	0,0226	0,0255	0,0261	0,0304	0,0495	0,0145	0,0171	0,0213	0,0255	0,0321	0,0541
0,79	0,0085	0,0108	0,0128	0,0133	0,0155	0,0252	0,0072	0,0085	0,0108	0,0128	0,0163	0,0277
0,795	0,0049	0,0059	0,0065	0,0066	0,0077	0,0186	0,0038	0,0043	0,0054	0,0064	0,0082	0,0139
0,798	0,0040	0,0055	0,0026	0,0047	0,0025	0,0048	0,0014	0,0017	0,0039	0,0026	0,0034	0,0057
0,8	0	0	0	0	0	0	0	0	0	0	0	0

τ*=0,8; **funct.**VIII; ψ=90° $I^{(1)}$ (τ, θ, ψ)

τ	0	15	30	45	60	75	0	15	30	45	60	75
0	0	0	0	0	0	0	0	0	0	0	0	0
0,2	0,0249	0,0274	0,0304	0,0345	0,0616	0,1235	0,0210	0,0214	0,0237	0,0337	0,0581	0,1106
0,4	0,0503	0,0553	0,0611	0,0674	0,1148	0,1981	0,0435	0,0441	0,0483	0,0677	0,1110	0,1819
0,5	0,0628	0,0689	0,0756	0,0820	0,1366	0,2201	0,0547	0,0552	0,0599	0,0832	0,1335	0,2038
0,6	0,0748	0,0820	0,0895	0,0951	0,1553	0,2341	0,0656	0,0659	0,0710	0,0977	0,1532	0,2181
0,7	0,1270	0,1328	0,1389	0,1494	0,2153	0,3096	0,1073	0,1083	0,1164	0,1475	0,2092	0,2945
0,75	0,1529	0,1580	0,1630	0,1752	0,2427	0,3403	0,1281	0,1296	0,1387	0,1717	0,2354	0,3264
0,78	0,1683	0,1730	0,1773	0,1903	0,2584	0,3566	0,1406	0,1422	0,1520	0,1859	0,2504	0,3438
0,79	0,1735	0,1779	0,1821	0,1952	0,2635	0,3618	0,1449	0,1464	0,1565	0,1906	0,2554	0,3495
0,795	0,1761	0,1805	0,1844	0,1977	0,2660	0,3644	0,1470	0,1486	0,1587	0,1930	0,2580	0,3523
0,798	0,1775	0,1819	0,1858	0,1992	0,2676	0,3659	0,1482	0,1499	0,1600	0,1944	0,2595	0,3538
0,8	0,1786	0,1829	0,1867	0,2002	0,2686	0,3669	0,1490	0,1507	0,1609	0,1954	0,2604	0,3549

$I^{(2)}$ (τ, θ, ψ)

τ	0	15	30	45	60	75	0	15	30	45	60	75
0	0,7822	0,7126	0,6080	0,4979	0,3865	0,4298	0,4288	0,4169	0,3835	0,3568	0,3502	0,3620
0,2	0,6307	0,5744	0,4916	0,4124	0,3651	0,3091	0,3551	0,3460	0,3203	0,3056	0,3101	0,3520
0,4	0,4226	0,3865	0,3388	0,2977	0,2836	0,3554	0,2537	0,2487	0,2359	0,2321	0,2518	0,3306
0,5	0,3004	0,2797	0,2515	0,2335	0,2375	0,3177	0,1938	0,1917	0,1869	0,1908	0,2153	0,3033
0,6	0,1612	0,1566	0,1543	0,1636	0,1889	0,2721	0,1259	0,1274	0,1327	0,1446	0,1744	0,2676
0,7	0,0873	0,0856	0,0835	0,0893	0,1055	0,1671	0,0685	0,0695	0,0724	0,0795	0,0981	0,1634
0,75	0,0451	0,0436	0,0433	0,0467	0,0555	0,0916	0,0355	0,0359	0,0378	0,0415	0,0518	0,0899
0,78	0,0183	0,0176	0,0176	0,0191	0,0228	0,0386	0,0145	0,0147	0,0155	0,0171	0,0213	0,0379
0,79	0,0085	0,0084	0,0088	0,0097	0,0116	0,0197	0,0072	0,0073	0,0078	0,0085	0,0108	0,0194
0,795	0,0049	0,0046	0,0044	0,0048	0,0057	0,0158	0,0038	0,0037	0,0040	0,0043	0,0055	0,0097
0,798	0,0040	0,0042	0,0019	0,0034	0,0019	0,0038	0,0014	0,0014	0,0028	0,0017	0,0022	0,0040
0,8	0	0	0	0	0	0	0	0	0	0	0	0

Table I 95

TABLE I (continued)

	60						75					ζ, °
0	15	30	45	60	75	0	15	30	45	60	75	θ,° / τ
0	0	0	0	0	0	0	0	0	0	0	0	0
0,0161	0,0194	0,0206	0,0340	0,0694	0,1645	0,0077	0,0083	0,0122	0,0197	0,0345	0,0889	0,2
0,0350	0,0422	0,0437	0,0706	0,1364	0,2779	0,0190	0,0201	0,0299	0,0462	0,0779	0,1785	0,4
0,0447	0,0541	0,0549	0,0879	0,1663	0,3163	0,0257	0,0270	0,0406	0,0620	0,1016	0,2237	0,5
0,0546	0,0663	0,0656	0,1042	0,1930	0,3411	0,0338	0,0348	0,0534	0,0805	0,1282	0,2723	0,6
0,0905	0,1006	0,1035	0,1474	0,2417	0,3997	0,0603	0,0625	0,0836	0,1166	0,1774	0,3300	0,7
0,1092	0,1182	0,1229	0,1692	0,2662	0,4291	0,0757	0,0785	0,1011	0,1377	0,2063	0,3679	0,75
0,1200	0,1290	0,1346	0,1825	0,2809	0,4468	0,0857	0,0890	0,1126	0,1516	0,2254	0,3941	0,78
0,1244	0,1327	0,1386	0,1869	0,2859	0,4528	0,0892	0,0926	0,1167	0,1564	0,2321	0,4035	0,79
0,1264	0,1345	0,1406	0,1891	0,2883	0,4558	0,0910	0,0945	0,1188	0,1588	0,2356	0,4083	0,795
0,1276	0,1355	0,1418	0,1905	0,2899	0,4577	0,0921	0,0956	0,1200	0,1604	0,2376	0,4113	0,798
0,1284	0,1363	0,1426	0,1913	0,2908	0,4588	0,0928	0,0963	0,1208	0,1613	0,2391	0,4133	0,8

60						75						ζ, °
0	15	30	45	60	75	0	15	30	45	60	75	θ,° / τ
0,2160	0,3015	0,4198	0,5506	0,6594	0,6805	0,1094	0,1236	0,1613	0,2233	0,2946	0,3162	0
0,1883	0,2641	0,3704	0,4864	0,6191	0,5879	0,1031	0,1177	0,1565	0,2237	0,3143	0,3276	0,2
0,1489	0,2033	0,2807	0,3776	0,4910	0,6504	0,0909	0,1041	0,1388	0,2003	0,2946	0,4328	0,4
0,1253	0,1636	0,2181	0,2902	0,3830	0,5435	0,0813	0,0931	0,1226	0,1729	0,2546	0,3987	0,5
0,0990	0,1162	0,1412	0,1802	0,2401	0,3697	0,0688	0,0785	0,0993	0,1308	0,1854	0,3027	0,6
0,0547	0,0648	0,0791	0,1020	0,1403	0,2348	0,0407	0,0466	0,0598	0,0800	0,1171	0,2100	0,7
0,0286	0,0340	0,0419	0,0543	0,0756	0,1321	0,0221	0,0254	0,0328	0,0442	0,0658	0,1231	0,75
0,0117	0,0140	0,0172	0,0224	0,0315	0,0566	0,0092	0,0107	0,0138	0,0187	0,0282	0,0542	0,78
0,0059	0,0070	0,0087	0,0114	0,0160	0,0289	0,0047	0,0054	0,0071	0,0096	0,0144	0,0280	0,79
0,0030	0,0036	0,0043	0,0057	0,0080	0,0145	0,0023	0,0028	0,0036	0,0048	0,0073	0,0142	0,795
0,0012	0,0014	0,0015	0,0023	0,0032	0,0059	0,0009	0,0011	0,0013	0,0019	0,0029	0,0057	0,798
0	0	0	0	0	0	0	0	0	0	0	0	0,8

60						75						ζ, °
0	15	30	45	60	75	0	15	30	45	60	75	θ,° / τ
0	0	0	0	0	0	0	0	0	0	0	0	0
0,0161	0,0172	0,0211	0,0290	0,0445	0,0857	0,0077	0,0082	0,0096	0,0129	0,0207	0,0430	0,2
0,0350	0,0375	0,0457	0,0613	0,0891	0,1490	0,0190	0,0200	0,0232	0,0300	0,0458	0,0834	0,4
0,0447	0,0478	0,0581	0,0773	0,1090	0,1705	0,0257	0,0271	0,0311	0,0396	0,0585	0,1003	0,5
0,0546	0,0585	0,0707	0,0930	0,1275	0,1865	0,0338	0,0355	0,0401	0,0502	0,0720	0,1169	0,6
0,0905	0,0952	0,1097	0,1366	0,1808	0,2634	0,0603	0,0628	0,0701	0,0855	0,1177	0,1853	0,7
0,1092	0,1141	0,1298	0,1587	0,2069	0,2975	0,0757	0,0785	0,0872	0,1057	0,1431	0,2213	0,75
0,1206	0,1257	0,1420	0,1721	0,2223	0,3168	0,0857	0,0889	0,0985	0,1188	0,1594	0,2439	0,78
0,1244	0,1296	0,1461	0,1765	0,2276	0,3232	0,0892	0,0925	0,1025	0,1233	0,1651	0,2518	0,79
0,1264	0,1316	0,1481	0,1788	0,2301	0,3264	0,0910	0,0943	0,1045	0,1256	0,1680	0,2557	0,795
0,1276	0,1327	0,1494	0,1800	0,2317	0,3283	0,0921	0,0954	0,1057	0,1270	0,1697	0,2581	0,798
0,1284	0,1335	0,1502	0,1810	0,2327	0,3295	0,0928	0,0961	0,1064	0,1279	0,1710	0,2597	0,8

60						75						ζ, °
0	15	30	45	60	75	0	15	30	45	60	75	θ,° / τ
0,2160	0,2164	0,2212	0,2370	0,2619	0,2634	0,1094	0,1107	0,1186	0,1310	0,1435	0,1363	0
0,1883	0,1891	0,1945	0,2039	0,2467	0,2169	0,1031	0,1051	0,1139	0,1291	0,1496	0,1286	0,2
0,1489	0,1500	0,1559	0,1743	0,2144	0,2808	0,0909	0,0930	0,1020	0,1186	0,1473	0,1914	0,4
0,1253	0,1267	0,1327	0,1497	0,1885	0,2689	0,0813	0,0832	0,0918	0,1080	0,1384	0,1979	0,5
0,0990	0,1007	0,1070	0,1218	0,1564	0,2491	0,0688	0,0703	0,0779	0,0928	0,1236	0,1989	0,6
0,0547	0,0558	0,0594	0,0682	0,0900	0,1559	0,0407	0,0415	0,0463	0,0558	0,0766	0,1349	0,7
0,0286	0,0292	0,0312	0,0360	0,0480	0,0869	0,0221	0,0225	0,0252	0,0305	0,0425	0,0784	0,75
0,0117	0,0120	0,0128	0,0148	0,0199	0,0370	0,0092	0,0095	0,0106	0,0128	0,0181	0,0343	0,78
0,0059	0,0060	0,0065	0,0075	0,0100	0,0189	0,0047	0,0048	0,0054	0,0066	0,0092	0,0176	0,79
0,0030	0,0031	0,0032	0,0037	0,0051	0,0094	0,0023	0,0024	0,0027	0,0033	0,0047	0,0089	0,795
0,0012	0,0012	0,0011	0,0015	0,0020	0,0039	0,0009	0,0009	0,0010	0,0013	0,0019	0,0036	0,798
0	0	0	0	0	0	0	0	0	0	0	0	0,8

$\tau^* = 0,8;$ funct. VIII; $\psi = 135°$ $I^{(1)}(\tau, \theta, \psi)$

ζ,°	30						45					
τ \ θ,°	0	15	30	45	60	75	0	15	30	45	60	75
0	0	0	0	0	0	0	0	0	0	0	0	0
0,2	0,0249	0,0266	0,0235	0,0366	0,0615	0,0981	0,0210	0,0283	0,0255	0,0345	0,0576	0,0906
0,4	0,0503	0,0538	0,0471	0,0720	0,1151	0,1577	0,0435	0,0590	0,0521	0,0700	0,1113	0,1503
0,5	0,0628	0,0669	0,0578	0,0880	0,1367	0,1745	0,0547	0,0746	0,0655	0,0866	0,1343	0,1682
0,6	0,0748	0,0797	0,0674	0,1028	0,1575	0,1846	0,0656	0,0904	0,0783	0,1023	0,1557	0,1796
0,7	0,1270	0,1378	0,1337	0,1722	0,2283	0,2881	0,1073	0,1368	0,1357	0,1683	0,2288	0,2859
0,75	0,1529	0,1667	0,1658	0,2055	0,2609	0,3299	0,1281	0,1603	0,1644	0,2007	0,2633	0,3301
0,78	0,1683	0,1839	0,1851	0,2251	0,2797	0,3522	0,1406	0,1743	0,1815	0,2199	0,2834	0,3542
0,79	0,1735	0,1896	0,1914	0,2316	0,2859	0,3592	0,1449	0,1791	0,1872	0,2263	0,2900	0,3617
0,795	0,1761	0,1925	0,1946	0,2349	0,2888	0,3627	0,1470	0,1814	0,1902	0,2296	0,2934	0,3657
0,798	0,1775	0,1942	0,1965	0,2368	0,2907	0,3648	0,1482	0,1829	0,1918	0,2314	0,2954	0,3679
0,8	0,1786	0,1954	0,1977	0,2380	0,2919	0,3662	0,1490	0,1838	0,1929	0,2327	0,2966	0,3673

$I^{(2)}(\tau, \psi, \theta)$

τ	0	15	30	45	60	75	0	15	30	45	60	75
0	0,7822	0,5374	0,4015	0,3282	0,2776	0,3220	0,4288	0,2928	0,2564	0,2493	0,2351	0,2842
0,2	0,6307	0,4311	0,3233	0,2720	0,2774	0,2281	0,3551	0,2424	0,2138	0,2135	0,2108	0,2229
0,4	0,4226	0,2950	0,2313	0,2035	0,2220	0,2833	0,2537	0,1805	0,1628	0,1693	0,1839	0,2814
0,5	0,3004	0,2208	0,1829	0,1685	0,1935	0,2691	0,1938	0,1476	0,1356	0,1450	0,1692	0,2706
0,6	0,1611	0,1377	0,1313	0,1324	0,1638	0,2562	0,1259	0,1131	0,1072	0,1188	0,1551	0,2563
0,7	0,0873	0,0750	0,0707	0,0718	0,0911	0,1576	0,0685	0,0615	0,0581	0,0650	0,0871	0,1563
0,75	0,0451	0,0382	0,0365	0,0373	0,0478	0,0862	0,0355	0,0317	0,0301	0,0338	0,0458	0,0862
0,78	0,0183	0,0154	0,0149	0,0153	0,0196	0,0363	0,0145	0,0130	0,0123	0,0188	0,0189	0,0365
0,79	0,0085	0,0074	0,0074	0,0078	0,0100	0,0185	0,0072	0,0064	0,0062	0,0069	0,0096	0,0186
0,795	0,0049	0,0040	0,0038	0,0038	0,0047	0,0152	0,0038	0,0033	0,0032	0,0035	0,0048	0,0093
0,798	0,0040	0,0037	0,0016	0,0026	0,0017	0,0036	0,0014	0,0013	0,0021	0,0014	0,0020	0,0038
0,8	0	0	0	0	0	0	0	0	0	0	0	0

$\tau^* = 0,8;$ funct. VIII; $\psi = 180°$ $I^{(1)}\tau, \theta, \psi)$

τ	0	15	30	45	60	75	0	15	30	45	60	75
0	0	0	0	0	0	0	0	0	0	0	0	0
0,2	0,0249	0,0296	0,0347	0,0325	0,0614	0,0974	0,0210	0,0286	0,0231	0,0425	0,0536	0,0932
0,4	0,0503	0,0601	0,0701	0,0637	0,1153	0,1578	0,0435	0,0596	0,0473	0,0864	0,1037	0,1564
0,5	0,0628	0,0750	0,0872	0,0777	0,1382	0,1755	0,0547	0,0754	0,0589	0,1078	0,1248	0,1766
0,6	0,0748	0,0897	0,1039	0,0903	0,1587	0,1870	0,0656	0,0915	0,0700	0,1288	0,1444	0,1910
0,7	0,1270	0,1470	0,1614	0,1678	0,2457	0,2994	0,1073	0,1421	0,1252	0,1936	0,2389	0,3301
0,75	0,1529	0,1755	0,1899	0,2052	0,2861	0,3448	0,1281	0,1679	0,1653	0,2255	0,2838	0,3884
0,78	0,1683	0,1926	0,2068	0,2271	0,3094	0,3693	0,1406	0,1832	0,1842	0,2446	0,3100	0,4206
0,79	0,1735	0,1982	0,2124	0,2344	0,3171	0,3771	0,1449	0,1884	0,1907	0,2508	0,3187	0,4309
0,795	0,1761	0,2011	0,2152	0,2380	0,3209	0,3808	0,1470	0,1911	0,1939	0,2540	0,3229	0,4360
0,798	0,1775	0,2027	0,2169	0,2402	0,3231	0,3830	0,1482	0,1926	0,1957	0,2559	0,3256	0,4389
0,8	0,1786	0,2039	0,2179	0,2416	0,3246	0,3845	0,1490	0,1936	0,1969	0,2571	0,3273	0,4409

$I^{(2)}(\tau, \theta, \psi)$

τ	0	15	30	45	60	75	0	15	30	45	60	75
0	0,7822	0,4863	0,3554	0,3532	0,2490	0,3387	0,4288	0,2570	0,2484	0,2245	0,2592	0,2994
0,2	0,6307	0,3893	0,2843	0,2653	0,2470	0,2434	0,3551	0,2120	0,2063	0,1869	0,2283	0,2245
0,4	0,4226	0,2683	0,2039	0,1963	0,2014	0,2951	0,2537	0,1598	0,1561	0,1496	0,1917	0,2880
0,5	0,3004	0,2036	0,1638	0,1612	0,1795	0,2766	0,1938	0,1337	0,1294	0,1315	0,1724	0,2785
0,6	0,1612	0,1326	0,1227	0,1255	0,1601	0,2560	0,1259	0,1075	0,1016	0,1160	0,1542	0,2709
0,7	0,0873	0,0721	0,0660	0,0678	0,0891	0,1575	0,0685	0,0584	0,0549	0,0633	0,0867	0,1657
0,75	0,0451	0,0367	0,0341	0,0352	0,0467	0,0863	0,0355	0,0301	0,0284	0,0329	0,0456	0,0916
0,78	0,0183	0,0148	0,0138	0,0144	0,0192	0,0364	0,0145	0,0123	0,0116	0,0134	0,0188	0,0388
0,79	0,0085	0,0071	0,0070	0,0073	0,0098	0,0184	0,0072	0,0061	0,0058	0,0067	0,0095	0,0198
0,795	0,0049	0,0038	0,0035	0,0036	0,0048	0,0153	0,0038	0,0031	0,0030	0,0033	0,0048	0,0098
0,798	0,0040	0,0035	0,0014	0,0024	0,0015	0,0036	0,0014	0,0012	0,0020	0,0014	0,0019	0,0041
0,8	0	0	0	0	0	0	0	0	0	0	0	0

Table I   97

TABLE I (continued)

60						75						ζ,°
θ,° 0	15	30	45	60	75	0	15	30	45	60	75	τ
0	0	0	0	0	0	0	0	0	0	0	0	0
0,0161	0,0157	0,0213	0,0291	0,0423	0,0814	0,0077	0,0079	0,0081	0,0103	0,0194	0,0418	0,2
0,0350	0,0340	0,0464	0,0621	0,0857	0,1432	0,0190	0,0194	0,0189	0,0247	0,0433	0,0818	0,4
0,0447	0,0432	0,0595	0,0787	0,1055	0,1647	0,0257	0,0262	0,0249	0,0324	0,0554	0,0987	0,5
0,0546	0,0525	0,0729	0,0957	0,1240	0,1815	0,0338	0,0342	0,0314	0,0405	0,0686	0,1155	0,6
0,0905	0,0918	0,1176	0,1506	0,1949	0,2809	0,0603	0,0628	0,0650	0,0819	0,1237	0,2004	0,7
0,1092	0,1121	0,1408	0,1787	0,2298	0,3253	0,0757	0,0793	0,0843	0,1056	0,1546	0,2451	0,75
0,1206	0,1246	0,1549	0,1959	0,2504	0,3504	0,0857	0,0902	0,0969	0,1210	0,1746	0,2733	0,78
0,1244	0,1287	0,1596	0,2016	0,2574	9,3587	0,0892	0,0939	0,1013	0,1264	0,1816	0,2831	0,79
0,1264	0,1307	0,1620	0,2044	0,2609	0,3626	0,0910	0,0959	0,1036	0,1291	0,1851	0,2880	0,795
0,1276	0,1321	0,1635	0,2061	0,2631	0,3653	0,0921	0,0970	0,1050	0,1308	0,1872	0,2910	0,798
0,1284	0,1330	0,1644	0,2073	0,2644	0,3669	0,0928	0,0978	0,1058	0,1319	0,1887	0,2930	0,8

60						75						
0,2160	0,1854	0,1741	0,1738	0,2186	0,2339	0,1074	0,0998	0,1020	0,1176	0,1318	0,1296	0
0,1883	0,1617	0,1534	0,1471	0,2059	0,2460	0,1031	0,0947	0,0978	0,1163	0,1380	0,1400	0,2
0,1489	0,1294	0,1257	0,1329	0,1832	0,2555	0,0909	0,0847	0,0891	0,1086	0,1388	0,1855	0,4
0,1253	0,1103	0,1097	0,1202	0,1661	0,2533	0,0813	0,0771	0,0822	0,1005	0,1337	0,1977	0,5
0,0990	0,0893	0,0926	0,1076	0,1466	0,2522	0,0688	0,0674	0,0735	0,0891	0,1255	0,2120	0,6
0,0547	0,0493	0,0511	0,0605	0,0842	0,1583	0,0407	0,0397	0,0435	0,0535	0,0779	0,1440	0,7
0,0286	0,0257	0,0267	0,0318	0,0449	0,0883	0,0221	0,0215	0,0237	0,0292	0,0433	0,0841	0,75
0,0117	0,0105	0,0110	0,0130	0,0186	0,0376	0,0092	0,0091	0,0099	0,0123	0,0184	0,0368	0,78
0,0059	0,0053	0,0055	0,0066	0,0094	0,0192	0,0047	0,0046	0,0051	0,0063	0,0094	0,0189	0,79
0,0030	0,0027	0,0028	0,0033	0,0047	0,0096	0,0023	0,0023	0,0026	0,0031	0,0048	0,0095	0,795
0,0012	0,0011	0,0010	0,0013	0,0019	0,0039	0,0009	0,0009	0,0009	0,0013	0,0019	0,0039	0,798
0	0	0	0	0	0	0	0	0	0	0	0	0,8

60						75						
0	0	0	0	0	0	0	0	0	0	0	0	0
0,0161	0,0168	0,0217	0,0269	0,0473	0,0734	0,0077	0,0075	0,0080	0,0101	0,0183	0,0428	0,2
0,0350	0,0364	0,0473	0,0573	0,0973	0,1307	0,0190	0,0181	0,0190	0,0251	0,0407	0,0871	0,4
0,0447	0,0465	0,0607	0,0726	0,1213	0,1509	0,0257	0,0243	0,0252	0,0332	0,0526	0,1089	0,5
0,0546	0,0568	0,0747	0,0880	0,1454	0,1678	0,0338	0,0324	0,0321	0,0422	0,0642	0,1339	0,6
0,0905	0,0972	0,1230	0,1568	0,2243	0,3135	0,0603	0,0608	0,0677	0,0936	0,1380	0,2358	0,7
0,1092	0,1182	0,1557	0,1922	0,2637	0,3786	0,0757	0,0778	0,0888	0,1235	0,1799	0,2906	0,75
0,1206	0,1311	0,1728	0,2137	0,2872	0,4156	0,0857	0,0890	0,1018	0,1432	0,2071	0,3255	0,78
0,1244	0,1354	0,1785	0,2210	0,2952	0,4279	0,0892	0,0930	0,1066	0,1500	0,2167	0,3377	0,79
0,1264	0,1376	0,1814	0,2247	0,2992	0,4339	0,0910	0,0950	0,1090	0,1535	0,2215	0,3439	0,795
0,1276	0,1388	0,1832	0,2268	0,3017	0,4377	0,0921	0,0962	0,1105	0,1556	0,2244	0,3476	0,798
0,1284	0,1398	0,1844	0,2283	0,3032	0,4399	0,0928	0,0970	0,1114	0,1570	0,2264	0,3501	0,8

0

60						75						
0,2160	0,1787	0,1533	0,1659	0,1931	0,2045	0,1094	0,0956	0,1035	0,1449	0,1287	0,1280	0
0,1883	0,1563	0,1356	0,1422	0,1872	0,1752	0,1031	0,0906	0,0998	0,1148	0,1376	0,1312	0,2
0,1489	0,1252	0,1139	0,1312	0,1765	0,2507	0,0909	0,0816	0,0911	0,1096	0,1437	0,2112	0,4
0,1253	0,1069	0,1024	0,1195	0,1670	0,2620	0,0813	0,0749	0,0836	0,1035	0,1423	0,2376	0,5
0,0990	0,0865	0,0907	0,1072	0,1558	0,2763	0,0688	0,0667	0,0736	0,0945	0,1381	0,2682	0,6
0,0547	0,0476	0,0500	0,0602	0,0899	0,1742	0,0407	0,0392	0,0436	0,0570	0,0861	0,1847	0,7
0,0286	0,0248	0,0261	0,0316	0,0479	0,0976	0,0221	0,0212	0,0237	0,0312	0,0480	0,1088	0,75
0,0117	0,0101	0,0107	0,0130	0,0199	0,0417	0,0092	0,0090	0,0099	0,0131	0,0205	0,0479	0,78
0,0059	0,0051	0,0054	0,0065	0,0100	0,0213	0,0047	0,0045	0,0051	0,0067	0,0104	0,0246	0,79
0,0030	0,0026	0,0027	0,0033	0,0050	0,0107	0,0023	0,0023	0,0026	0,0034	0,0053	0,0125	0,795
0,0012	0,0010	0,0009	0,0013	0,0020	0,0043	0,0009	0,0009	0,0009	0,0014	0,0021	0,0051	0,798
0	0	0	0	0	0	0	0	0	0	0	0	0,8

$\tau^* = 0,2;$ funct. V

TABLE II

θ,° τ	$B_0^{(1)}(\tau, \theta)$						$B_0^{(2)}(\tau, \theta)$					
	0	15	30	45	60	75	0	15	30	45	60	75
0	1	1	1	1	1	1	0,0711	0,0738	0,0826	0,1021	0,1440	0,2662
0,05	0,9837	0,9832	0,9809	0,9759	0,9645	0,9286	0,0529	0,0550	0,0616	0,0763	0,1086	0,1988
0,1	0,9668	0,9655	0,9610	0,9612	0,9291	0,8630	0,0354	0,0368	0,0412	0,0511	0,0732	0,1379
0,125	0,9583	0,9566	0,9511	0,9389	0,9118	0,8329	0,0270	0,0281	0,0315	0,0391	0,0560	0,1032
0,15	0,9496	0,9476	0,9410	0.9265	0,8944	0,8039	0,0189	0,0196	0,0219	0,0271	0,0388	0,0743
0,16	0,9454	0,9434	0,9363	0,9208	0,8866	0,7917	0,0149	0,0156	0,0174	0,0216	0,0310	0,0599
0,17	0,9412	0,9389	9314	0,9148	0,8788	0,7795	0,0112	0,0116	0,0130	0,0161	0,0232	0,0453
0,18	0,9370	0,9346	0,9266	0,9090	0,8710	0,7675	0,0074	0,0077	0,0086	0,0107	0,0155	0.0305
0,19	0,9328	0,9303	0,9217	0,9032	0,8631	0,7558	0,0037	0,0038	0,0043	0.0053	0,0077	0,0153
0,195	0,9306	0,9280	0,9193	0,9004	0,8593	0,7499	0,0018	0,0019	0,0022	0,0027	0,0039	0,0077
0,197	0,9298	0,9271	0,9182	0,8991	0,8578	0,7477	0,0011	0,0011	0,0013	0,0016	0,0023	0,0046
0,200	0,9285	0,9258	0,9168	0,8974	0,8554	0,7442	0	0	0	0	0	0

$\tau^* = 0,2;$ funct. VI

τ	0	15	30	45	60	75	0	15	30	45	60	75
0	1	1	1	1	1	1	0,0618	0,0645	0,0738	0,0912	0,1520	0,2432
0,05	0,9868	0,9862	0,9841	0,9796	0,9688	0,933	0,0470	0,0489	0,0563	0,0694	0,1008	0,1906
0,1	0,9730	0,9883	0,9677	0,9586	0,9376	0,872	0,0326	0,0339	0,0394	0,0479	0,0696	0,1344
0,125	0,9660	0,9645	0,9594	0,9482	0,9223	0,844	0,0257	0,0268	0,0313	0,0376	0,0544	0,1020
0,15	0,9589	0,9571	0,9510	0.9376	0,9070	0,8176	0,0191	0,0198	0,0234	0,0275	0,0393	0,0753
0,16	0,9547	0.9528	0,9462	0,9318	0,8990	0,8051	0,0152	0,0158	0,0177	0,0219	0,0314	0,0608
0,17	0,9506	0,9483	0,9413	0,9258	0,8912	0,7926	0,0113	0,0118	0,0132	0,0164	0,0235	0,0459
0,18	0,9462	0,9440	0.9365	0,9200	0,8832	0,7804	0,0075	0,0078	0,0088	0,0109	0,0157	0,0309
0,19	0,9420	0,9396	0,9315	0,9140	0,8752	0,7684	0,0038	0,0039	0,0043	0,0054	0,0078	0,0155
0,195	0,9398	0,9373	0,9291	0,9112	0,8713	0,7625	0,0019	0,0019	0,0022	0,0027	0,0039	0,0078
0,197	0,9390	0,9365	0,9281	0,9099	0,8699	0,7601	0,0011	0,0012	0,0013	0,0016	0,0023	0,0047
0,200	0,9377	0,9352	0,9266	0,9082	0,8674	0,7566	0	0	0	0	0	0

$\tau^* = 0,4;$ funct. VI

τ	0	15	30	45	60	75	0	15	30	45	60	75
0	1,0000	1,0000	1,0000	1,0000	1,0000	1,0000	0,1190	0,1236	0,1385	0,1700	0,2348	0,3824
0,1	0,9745	0,9734	0,9696	0,9613	0,9419	0,8825	0,0889	0,0924	0,1036	0,1277	0,1789	0,3047
0,2	0,9469	0,9446	0,9370	0,9204	0,8527	0,7817	0,0604	0,0627	0,0704	0,0870	0,1229	0,2190
0,25	0,9332	0,9304	0,9210	0,9006	0,8561	0,7401	0,0475	0,0492	0,0552	0,0681	0,0964	0,1752
0,3	0,9191	0,9157	0,9045	0,8805	0,8290	0,7008	0,0348	0,0360	0,0403	0,0494	0,0695	0,1274
0,32	0,9109	0,9073	0,8953	0,8695	0,8151	0,6825	0,0276	0,0286	0,0320	0,0393	0,0556	0,1036
0,34	0,9027	0,8987	0,8859	0,8586	0,8012	0,6645	0,0205	0,0213	0,0238	0,0293	0,0416	0,0789
0,36	0,8943	0,8902	0,8764	0,8474	0,7872	0,6468	0,0136	0,0141	0,0157	0,0194	0,0277	0,0534
0,38	0,8858	0,8815	0,8670	0,8364	0,7733	0,6294	0,0067	0,0069	0,0078	0,0095	0,0138	0,0270
0,39	0,8816	0,8772	0,8622	0,8309	0,7665	0,6208	0,0033	0,0034	0,0038	0,0048	0,0069	0,0136
0,395	0,8795	0,8750	0,8598	0,8282	0,7630	0,6167	0,0017	0,0017	0,0019	0,0024	0,0034	0,0068
0,397	0,8788	0,8742	0,8589	0,8269	0,7616	0,6150	0,0010	0,0010	0,0012	0,0014	0,0021	0,0041
0,4	0,8774	0,8727	0,8574	0,8253	0,7596	0,6125	0	0	0	0	0	0

$\tau^* = 0,4;$ funct. VII

τ	0	15	30	45	60	75	0	15	30	45	60	75
0	1,000	1,000	1,000	1,000	1,000	1,000	0,0928	0,0970	0,1105	0,1401	0,2035	0,3547
0,1	0,9837	0,9829	0,9798	0,9728	0,9554	0,8983	0,0726	0,0758	0,0862	0,1092	0,1597	0,2895
0,2	0,9655	0,9637	0,9574	0,9430	0,8742	0,8085	0,0533	0,0554	0,0627	0,0787	0,1144	0,2130
0,25	0,9562	0,9542	0,9464	0,9287	0,8872	0,7721	0,0446	0,0464	0,0521	0,0650	0,0935	0,1745
0,3	0,9471	0,9444	0,9351	0,9140	0,8658	0,7380	0,0362	0,0375	0,0419	0,0515	0,0724	0,1329
0,32	0,9387	0,9358	0,9256	0,9027	0,8512	0,7184	0,0288	0,0298	0,0333	0,0409	0,0579	0,1081
0,34	0,9304	0,9270	0,9160	0,8913	0,8366	0,6992	0,0214	0,0222	0,0248	0,0306	0,0434	0,0823
0,36	0,9218	0,9184	0,9063	0,8798	0,8220	0,6802	0,0141	0,0146	0,0164	0,0202	0,0289	0,0557
0,38	0,9131	0,9094	0,8966	0,8684	0,8074	0,6616	0,0070	0,0072	0,0081	0,0100	0,0144	0,0282
0,39	0,9088	0,9050	0,8917	0,8626	0,8003	0,6525	0,0035	0,0036	0,0040	0,0050	0,0072	0,0142
0,395	0,9068	0,9028	0,8892	0,8598	0,7968	0,6481	0,0017	0,0018	0,0020	0,0025	0,0036	0,0071
0,397	0,9060	0,9019	0,8882	0,8586	0,7953	0,6463	0,0010	0,0011	0,0012	0,0015	0,0022	0,0043
0,4	0,9046	0.9005	0,8867	0,8568	0,7931	0,6436	0	0	0	0	0	0

Table II 99

$\tau^* = 0,6$; funct. VII

TABLE II (continued)

θ.° τ	$B_0^{(1)}(\tau, \theta)$						$B_0^{(2)}(\tau, \theta)$					
	0	15	30	45	60	75	0	15	30	45	60	75
0	1,0000	1,0000	1,0000	1,0000	1,0000	1,0000	0,1357	0,1414	0,1595	0,1980	0,2750	0,6067
0,1	0,9848	0,9840	0,9812	0,9747	0,9586	0,9057	0,1239	0,1291	0,1456	0,1810	0,2530	0,5658
0,2	0,9683	0,9667	0,9611	0,9482	0,9175	0,8278	0,1052	0,1094	0,1236	0,1541	0,2177	0,4998
0,3	0,9508	0,9484	0,9398	0,9207	0,8771	0,7618	0,0870	0,0905	0,1022	0,1275	0,1819	0,4231
0,4	0,9323	0,9290	0,9176	0,8927	0,8376	0,7049	0,0696	0,0723	0,0815	0,1015	0,1456	0,3290
0,45	0,9158	0,9119	0,8976	0,8664	0,7977	0,6287	0,0504	0,0521	0,0580	0,0706	0,0979	0,1724
0,50	0,8948	0,8902	0,8740	0,8382	0,7647	0,5941	0,0329	0,0340	0,0379	0,0464	0,0651	0,1191
0,55	0,8735	0,8684	0,8502	0,8116	0,7338	0,5601	0,0159	0,0165	0,0184	0,0226	0,0321	0,0612
0,57	0,8650	0,8597	0,8407	0,8009	0,7192	0,5472	0,0094	0,0098	0,0109	0,0134	0,0191	0,0372
0,58	0,8608	0,8553	0,8361	0,7954	0,7130	0,5409	0,0063	0,0066	0,0072	0,0089	0,0128	0,0250
0,59	0,8566	0,8509	0,8313	0,7901	0,7067	0,5347	0,0031	0,0032	0,0036	0,0044	0,0064	0,0126
0,595	0,8544	0,8488	0,8289	0,7874	0,7036	0,5311	0,0016	0,0016	0,0018	0,0022	0,0032	0,0063
0,6	0,8523	0,8467	0,8266	0,7848	0,7004	0,5284	0	0	0	0	0	0

$\tau^* = 0,6$; funct. VIII

θ.° τ												
0	1	1	1	1	1	1	0,1095	0,1146	0,1306	0,1661	0,2414	0,4022
0,1	0,9906	0,9899	0,9877	0,9823	0,9678	0,9167	0,1024	0,1071	0,1220	0,1551	0,2260	0,3815
0,2	0,9802	0,9788	0,9744	0,9636	0,9357	0,8483	0,0899	0,0940	0,1069	0,1357	0,1987	0,3456
0,3	0,9691	0,9670	0,9603	0,9440	0,9041	0,7910	0,0781	0,0814	0,0923	0,1168	0,1712	0,3082
0,4	0,9573	0,9544	0,9454	0,9239	0,8732	0,7420	0,0665	0,0692	0,0781	0,0680	0,1428	0,2668
0,45	0,9436	0,9400	0,9281	0,9003	0,8352	0,6651	0,0527	0,0546	0,0607	0,0739	0,1027	0,1809
0,50	0,9223	0,9180	0,9041	0,8722	0,8007	0,6278	0,0344	0,0356	0,0397	0,0486	0,0682	0,1250
0,55	0,9006	0,8958	0,8796	0,8438	0,7683	0,5913	0,0166	0,0172	0,0192	0,0236	0,0336	0,0641
0,57	0,8920	0,8869	0,8700	0,8327	0,7531	0,5775	0,0099	0,0102	0,0114	0,0140	0,0201	0,0390
0,58	0,8877	0,8825	0,8651	0,8271	0,7465	0,5708	0,0065	0,0068	0,0074	0,0093	0,0134	0,0282
0,59	0,8833	0,8781	0,8603	0,8216	0,7399	0,5641	0,0033	0,0033	0,0037	0,0046	0,0067	0,0132
0,595	0,8812	0,8759	0,8579	0,8188	0,7367	0,5607	0,0016	0,0017	0,0019	0,0023	0,0034	0,0066
0,6	0,8791	0,8737	0,8555	0,8161	0,7334	0,5574	0	0	0	0	0	0

$\tau^* = 0,8$; funct. VII

θ.° τ												
0	1	1	1	1	1	1	0,1546	0,1611	0,2012	0,2445	0,3271	0,4791
0,2	0,9655	0,9637	0,9574	0,9432	0,9098	0,8144	0,1100	0,1148	0,1297	0,1882	0,2550	0,3555
0,4	0,9279	0,9244	0,9123	0,8860	0,8290	0,6956	0,0691	0,0719	0,0815	0,1379	0,1905	0,3093
0,5	0,9092	0,9051	0,8904	0,8591	0,7935	0,6521	0,0496	0,0516	0,0931	0,1138	0,1573	0,2627
0,6	0,8899	0,8850	0,8679	0,7917	0,7587	0,6119	0,0306	0,0316	0,0352	0,0900	0,1233	0,2095
0,7	0,8481	0,8421	0,8215	0,7790	0,6968	0,5435	0,0227	0,0235	0,0262	0,0430	0,0603	0,1103
0,75	0,8275	0,8209	0,7988	0,7535	0,6673	0,5129	0,0148	0,0154	0,0171	0,0210	0,0298	0,0568
0,78	0,8151	0,8083	0,7852	0,7382	0,6501	0,4953	0,0058	0,0060	0,0067	0,0082	0,0118	0,0231
0,79	0,8110	0,8040	0,7808	0,7332	0,6445	0,4896	0,0029	0,0030	0,0033	0,0041	0,0059	0,0116
0,795	0,8090	0,8019	0,7785	0,7307	0,6417	0,4869	0,0014	0,0015	0,0017	0,0021	0,0030	0,0058
0,798	0,8076	0,8007	0,7771	0,7293	0,6400	0,4852	0,0006	0,0006	0,0008	0,0008	0,0012	0,0023
0,8	0,8069	0,7998	0,7763	0,7283	0,6389	0,4840	0	0	0	0	0	0

$\tau^* = 0,8$; funct. VIII

θ.° τ												
0	1	1	1	1	1	1	0,1423	0,1485	0,1679	0,2081	0,2893	0,4459
0,2	0,9782	0,9765	0,9716	0,9595	0,9288	0,8349	0,1152	0,1200	0,1355	0,1674	0,2334	0,3773
0,4	0,9539	0,9508	0,9416	0,9183	0,8654	0,7333	0,0910	0,0944	0,1066	0,1312	0,1847	0,3087
0,5	0,9417	0,9380	0,9209	0,8987	0,8375	0,6965	0,0798	0,0827	0,0931	0,1135	0,1586	0,2694
0,6	0,9294	0,9246	0,9108	0,8790	0,8103	0,6627	0,0691	0,0714	0,0800	0,0961	0,1318	0,2242
0,7	0,8866	0,8806	0,8628	0,8237	0,7440	0,5868	0,0325	0,0336	0,0375	0,0459	0,0644	0,1178
0,75	0,8654	0,8588	0,8393	0,7969	0,7127	0,5531	0,0158	0,0163	0,0182	0,0224	0,0318	0,0607
0,78	0,8525	0,8457	0,8252	0,7809	0,6942	0,5337	0,0062	0,0064	0,0071	0,0088	0,0126	0,0246
0,79	0,8484	0,8414	0,8205	0,7757	0,6882	0,5274	0,0031	0,0032	0,0035	0,0044	0,0063	0,0124
0,795	0,8463	0,8392	0,8182	0,7730	0,6852	0,5244	0,0015	0,0016	0,0018	0,0022	0,0031	0,0062
0,798	0,8449	0,8380	0,8168	0,7715	0,6834	0,5225	0,0006	0,0006	0,0008	0,0009	0,0013	0,0025
0,8	0,8441	0,8371	0,8158	0,7704	0,6821	0,5213	0	0	0	0	0	0

TABLE III

$\tau^* = 0,2; \frac{1}{2} C \cdot e^{-\tau^* \sec \zeta}$

No. funct.	V				VI			
q \ $\zeta,°$	30	45	60	75	30	45	60	75
0,1	0,1598	0,1274	0,0852	0,0375	0,1616	0,1290	0,0863	0,0380
0,2	0,3241	0,2583	0,1727	0,0759	0,3273	0,2611	0,1748	0,0769
0,3	0,4926	0,3927	0,2626	0,1155	0,4970	0,3965	0,2654	0,1168
0,4	0,6658	0,5306	0,3548	0,1561	0,6711	0,5354	0,3584	0,1577
0,6	1,0267	0,8183	0,5472	0,2406	1,0325	0,8237	0,5514	0,2426
0,8	1,4083	1,1224	0,7506	0,3301	1,4132	1,1275	0,7547	0,3320

$\tau^* = 0,4; \frac{1}{2} C \cdot e^{-\tau^* \sec \zeta}$

	VI				VII			
0,1	0,1505	0,1175	0,0749	0,0296	0,1558	0,1218	0,0777	0,0306
0,2	0,3077	0,2400	0,1530	0,0604	0,3174	0,2481	0,1583	0,0623
0,3	0,4715	0,3678	0,2344	0,0926	0,4848	0,3789	0,2418	0,0951
0,4	0,6426	0,5013	0,3195	0,1262	0,6584	0,5163	0,3284	0,1292
0,6	1,0088	0,7871	0,5016	0,1981	0,9945	0,8022	0,5117	0,2013
0,8	1,5670	1,1006	0,7014	0,2770	1,4212	1,1125	0,7098	0,2792

$\tau^* = 0,6; \frac{1}{2} C \cdot e^{-\tau^* \sec \zeta}$

	VII				VIII			
0,1	0,1480	0,1095	0,0691	0,0253	0,1531	0,1174	0,0716	0,0259
0,2	0,3044	0,2251	0,1421	0,0521	0,3128	0,2399	0,1462	0,0530
0,3	0,4697	0,3474	0,2192	0,0804	0,4795	0,3677	0,2240	0,0812
0,4	0,6449	0,4770	0,3010	0,1104	0,6535	0,5012	0,3053	0,1107
0,6	1,0284	0,7607	0,4800	0,1760	1,0259	0,7869	0,4793	0,1738
0,8	1,4637	1,0826	0,6832	0,2505	1,4348	1,1005	0,6704	0,2430

$\tau^* = 0,8; \frac{1}{2} C \cdot e^{-\tau^* \sec \zeta}$

	VII				VIII			
0,1	0,1386	0,0995	0,0606	0,0209	0,1454	0,1090	0,0634	0,0216
0,2	0,2854	0,2049	0,1248	0,0431	0,2982	0,2237	0,1302	0,0443
0,3	0,4410	0,3166	0,1928	0,0666	0,4595	0,3445	0,2004	0,0682
0,4	0,6064	0,4353	0,2651	0,0916	0,6293	0,4718	0,2745	0,0934
0,6	0,9701	0,6964	0,4241	0,1465	0,9984	0,7486	0,4356	0,1482
0,8	1,3856	0,9947	0,6057	0,2296	1,4126	1,0591	0,6162	0,2096

TABLE IV

τ^*	No. funct.	$F^{(2)}(0)$				$F^{(2)}(0) + 2\pi \cos\zeta \cdot e^{-\tau^* \sec\zeta}$				$\pi \int_0^{\pi/2} B_0^{(2)}(0,\theta)\sin 2\theta\, d\theta$
	$\zeta,°$	30	45	60	75	30	45	60	75	
0,2	V	0,6314	0,6043	0,5370	0,2330	4,9507	3,9526	2,6429	0,8980	0,3761
	VI	0,6901	0,6558	0,5738	0,2425	5,0094	4,0041	2,6797	0,9075	0,3752
0,4	VI	1,2073	1,0929	0,8937	0,5637	4,6359	3,6164	2,3053	0,8546	0,6414
	VII	1,3838	1,2376	0,9888	0,5973	4,8124	3,7611	2,4004	0,8882	0,2043
0,6	VII	1,8005	1,4471	1,1666	0,6150	4,5221	3,3488	2,1128	0,7423	0,8341
	VIII	1,9912	1,7133	1,2555	0,6652	4,7128	3,6150	2,2017	0,7925	0,6425
0,8	VII	2,0756	1,7437	1,0724	0,5656	4,2364	3,1775	1,7070	0,6391	0,8706
	VIII	2,2985	1,9069	1,3098	0,5876	4,4593	3,3407	1,9444	0,6611	0,7717

TABLE V

	$\tau^*=0,2$			$\tau^*=0,4$			$\tau^*=0,6$			$\tau^*=0,8$	
τ	V	VI	τ	VI	VII	τ	VII	VIII	τ	VII	VIII
0	0	0	0	0	0	0	0	0	0	0	0
0,05	1,8	0,9	0,1	1,8	0,9	0,1	0,9	0,7	0,2	1,8	1,5
0,10	4,5	2,2	0,2	4,4	2,2	0,2	2,0	1,6	0,4	4,5	3,6
0,125	6,3	3,1	0,25	6,3	3,2	0,3	3,4	2,7	0,5	6,3	5,2
0,15	8,9	4,4	0,3	8,9	4,5	0,4	5,3	4,3	0,6	8,9	7,3
0,16	10,4	6,2	0,32	10,3	5,2	0,45	6,7	5,5	0,7	13,4	10,3
0,17	12,2	6,1	0,34	12,1	6,2	0,5	8,7	7,1	0,75	17,9	14,6
0,18	14,8	7,4	0,36	14,7	7,4	0,55	12,0	9,8	0,78	23,8	19,4
0,19	19,3	9,6	0,38	19,2	9,7	0,57	14,5	11,8	0,79	28,3	23,1
0,195	23,8	11,8	0,39	23,6	11,9	0,58	16,4	13,4	0,795	32,7	26,7
0,197	27,1	13,4	0,395	28,0	14,1	0,59	19,8	16,2	0,798	38,6	31,5
0,2	∞	∞	0,397	31,3	15,8	0,595	23,2	18,9	0,8	∞	∞
			0,4	∞	∞	0,6	∞	∞			

REFERENCES

[1] E. S. Kuznetsov and B. V. Ovchinskii, "Results of a numerical solution of the integral equation of the theory of scattering of light in the atmosphere," Trudy Geofiz. Instituta, No. 4 (1949).

[2] E. S. Kuznetsov, "Theory of nonhorizontal visibility," Izv. AN SSSR, ser. geogr. i geofiz., No. 5 (1943).

[3] E. S. Kuznetsov, "Application of the formulas of the theory of nonhorizontal visibility to the calculation of the brightness of the sky and visibility range for simpler forms of scattering functions," Izv. AN SSSR, ser. geogr. i geofiz. No. 3 (1945).

[4] S. Chandrasekhar, The Transfer of Radiant Energy (IL, 1953).

[5] E. S. Kuznetsov, "A general method for setting up approximate equations of transfer of radiant energy," AN SSSR, ser. geofiz. No. 4 (1951).

[6] J. M. Waldram, "Measurements of the photometric properties of the upper atmosphere," Quart. J. Roy. Meteor. Soc. 71, N 309-310 (1945).

[7] J. M. Waldram, Symposium on Searchlights [London Illum. Eng. Soc. 1948].

[8] W. E. Middleton, Vision Through the Atmosphere (University of Toronto Press, 1952).

[9] V. F. Belov, Measurement of the Main Optical Characteristics of the Layer of the Atmosphere near the Earth's Surface. (Gidrometizdat, 1956).

[10] E. O. Hulburt, "Optics of atmospheric haze," J. Opt. Soc. Am. 31, 467-476 (1941).

[11] Y. Rocard, "Visibilité des buts par un projecteur," Revue d'Optique 11, 193-211 (1932).

[12] K. Bullrich, "Durchlässigheitszahl and Zerstreuungsfunktion in dunstiger Luft," Met. Z. 61, 270-273 (1944).

[13] K. Bullrich and F. Möller, "Die Streuung des Lichtes in trüber Luft," Optik. 2, 301-325 (1947).

[14] E. Reeger and H. Seidentopf, "Die Streufunktion des atmosphärischen Dunst nach Scheinwerfermessungen," Optik. 1, 15-41 (1946).

[15] L. Foitzik and Zschaeck, "Messungen der spektralen Zerstreuungsfunktion bodennaher Luft bei guter Sicht, Dunst, und Nebel," Zeitschr. f. Met. 7, 1, 1-19 (1952).

[16] E. V. Pyaskovskaya-Fesenkova, "Some data on the optical properties of the atmosphere in mountain conditions," Astron. Zh., Vol. XXIX, No. 3 (1952).

[17] E. V. Pyaskovskaya-Fesenkova, "Some data on the atmospheric light scattering function," Doklady AN SSSR, Vol. 86, No. 5 (1952).

[18] E. V. Pyaskovskaya-Fesenkova, "Some properties of atmospheric light scattering function," Doklady AN SSSR, Vol. 88, No. 1 (1953).

[19] S. D. Gutshabash, "The scattering of light in a medium with a variable scattering function," Uch. zap. LGU. Astronomiya, No. 25, (1952).

[20] M. S. Malkevich, "On the solution of integral equations of the theory of light scattering in the atmosphere," Izv. AN SSSR, ser. geofiz. No. 9 (1956).

[21] S. Ya. Kogan, "Application of spherical functions to the problem of light scattering in the atmosphere," Izv. AN SSSR ser. geofiz. No. 3 (1957).

[22] I. N. Yaroslavtsev, "Distribution of brightness over the sky," Izv. AN SSSR ser. geofiz. No. 1 (1953).

[23] K. Ya. Kondrat'ev, Radiant Energy of the Sun (Gidrometizdat, 1954).